Telenovelas and Transformation

This book investigates how telenovelas may be the key to the future of Brazilian television and how this content can survive in an interconnected media landscape.

Recognized telenovela writer and scholar Rosane Svartman considers the particular characteristics of the telenovela format – number of episodes, melodrama influence and influence of the audience on future writing – to explore how these can be preserved on multimedia platforms and the challenges this change may present. Svartman further charts the transformations of the telenovela throughout its history and its major influences and unveils the main storytelling elements and writing processes. Chapters examine the business model of Brazilian television within the current context of hypermedia and analyze how this relationship evolves as it is influenced by the new interactive tools and technologies that amplify the audience's power.

Merging empirical practices and theory, this book will be of great interest to scholars and students of transmedia storytelling, television studies and Latin American media, as well as professionals working in these areas.

Rosane Svartman has a PhD in Communication Studies/Cinema and is the head writer of four internationally recognized, award-nominee telenovelas. She wrote and directed five series for television and has five feature films. She is also the author of three theatre plays and has coordinated more than ten transmedia narrative projects.

She would also like to credit:

Theses supervisor: Felipe Muanis is Associate Professor at Cinema and Audiovisual Course at Federal University of Juiz de Fora, Brazil, and has published books about cinema and television image theory.
Copyediting: Roberta Rodrigues has been working as a translator and interpreter since 1993.

Routledge Advances in Transmedia Studies
Series Editor: Matthew Freeman

This series publishes monographs and edited collections that sit at the cutting-edge of today's interdisciplinary cross-platform media landscape. Topics should consider emerging transmedia applications in and across industries, cultures, arts, practices, or research methodologies. The series is especially interested in research exploring the future possibilities of an interconnected media landscape that looks beyond the field of media studies, notably broadening to include socio-political contexts, education, experience design, mixed-reality, journalism, the proliferation of screens, as well as art- and writing-based dimensions to do with the role of digital platforms like VR, apps and iDocs to tell new stories and express new ideas across multiple platforms in ways that join up with the social world.

Transmediality in Independent Journalism
The Turkish Case
Dilek Gürsoy

Theory, Development, and Strategy in Transmedia Storytelling
Edited by Renira Rampazzo Gambarato, Geane Carvalho Alzamora, Lorena Tárcia

Place and Immersion in Contemporary Transmedia Storytelling
Donna Hancox

Telenovelas and Transformation
Saving Brazil's Television Industry
Rosane Svartman

Telenovelas and Transformation
Saving Brazil's Television Industry

Rosane Svartman

Routledge
Taylor & Francis Group

LONDON AND NEW YORK

First published 2021
by Routledge
2 Park Square, Milton Park, Abingdon, Oxon OX14 4RN

and by Routledge
52 Vanderbilt Avenue, New York, NY 10017

Routledge is an imprint of the Taylor & Francis Group, an informa business

British Library Cataloguing-in-Publication Data
A catalogue record for this book is available from the British Library

Library of Congress Cataloging-in-Publication Data
A catalog record for this book has been requested

ISBN: 978-0-367-54368-6 (hbk)
ISBN: 978-1-003-08898-1 (ebk)

Typeset in Times New Roman
by Apex CoVantage, LLC

Contents

Figures

Acknowledgements

I want to express my most heartfelt gratitude and sincere thanks to my thesis supervisor, Felipe Muanis. He not only inspired me to go back to university for a PhD but also encouraged me along the way with his generosity and expertise. I received all the necessary help from Roberta Rodrigues with copyediting during the translation and adaptation of my thesis to English for this book. I also want to express my sincere appreciation to Eneida Nogueira and Miriam de Icaza Sánchez for sharing with me their experience as researchers. My thanks to professors Carla Barros, Lia Bahia, Maurício Bragrança, Tatiana Siciliano, Tunico Amâncio and Vera Lúcia Follain de Figueiredo for their support and advice. Also, I am thankful to Gisele Gomes and Beatriz Azeredo from Globo, for helping me with permissions. Finally, I am indebted to my family for their support and patience along the way.

1 Introduction

Nowadays, the boundaries between screen technologies have disintegrated. A film initially planned for a cinematographic device may be viewed on a mobile phone; a video made for a YouTube channel may also be seen on a connected television set. Even though the process of content slippage is no novelty, mainly when it comes to film and television, it is the proliferation of canvases, media and practices of viewer interaction and participation that results in the uniqueness of the current moment which, in turn, is accompanied by a transformation in spectatorship. We stand before an increasingly participant, multiplatform audience.

Marshall McLuhan (1988), published posthumously by his son Eric McLuhan, has observed that, over time, one medium would dominate another altogether. It remained for the old medium not to disappear but to transform into art. Consequently, film has replaced theatre, television has replaced film. This concept may be expanded in pondering whether, presently, television has not already been replaced as the dominant medium by the experience of connectivity and by the consumption of content on digital and interactive platforms. Even disagreeing with this position, it is impossible to ignore the transformations that new media have brought to spectatorship and the consumption of television. Television has changed systematically in a continuous historical process, both as a device and as a phenomenon. There are new potentialities that television gains in contemporaneity due to the advent of the digital and because of its uses.

Brazilian television, being the predominant medium of contemporaneity in the country, attracts millions of people daily to its programming. Currently, TV Globo, part of the largest media conglomerate in Latin America, is also the most significant television broadcaster in Brazil. In 2017, the broadcaster started a campaign among its audience after having celebrated a daily viewership of "over 100 million people" (2017). The fact that Brazil's television still has mass audiences to this day is a quintessential Brazilian phenomenon. In most countries, content, pulverized into various channels and platforms, also results in a fragmented audience. Digital platforms, such as Amazon, Netflix and Hulu, use precisely this audience segmentation, along with user data, to produce narratives with previously identified audience appeal. Netflix announced that in 2020 the platform totalled 167 million users worldwide in 190 countries. Nevertheless, none of the Netflix content has the priority to become a mass audience product, as we shall analyze further.

As a result of its generalist style, TV Globo, a public concession commercially exploited by a private group, creates what Dominique Wolton (1996) defines as a social link between Brazil's many realities, associating ideology and technique and creating various relationships, always in transformation. Also, in this sense, Esther Hamburger ponders: "Television has established itself as a medium able to address the most varied segments in social, age and regional terms" (2011). Muanis (2018) observes that television also provides a sense of up-to-date-ness that leads the audience to consume given content within a given deadline to participate in conversations and debates in one's social group. In a country of continental dimensions and distinct social and economic realities, the same news and narratives occupy the more significant part of programming. In this sense, television could also be an immense machine for influence. As noted by Barbosa (2010), the speech made 18 September 1951 by press mogul Assis Chateaubriand at the opening of Tupi TV, the first Brazilian television station, indicates the way this technology was perceived. In Chateaubriand's words, "the most subversive machine for influencing public opinion – a machine that gives flight to the most capricious fancy and brings together the most far-reaching groups of humanity" (Barbosa, 2010).

Built up over the last decades, TV Globo's vast linear programming viewership and economic power are widely criticized. Various groups accuse TV Globo's journalism of being tendentious, especially in its reporting of Brazil's political events to its broad audience. In the 2018 elections, both right- and left-wing politicians complained of TV Globo's election coverage. The question of the protagonist and narrative representativity in fictional content broadcasted is also a point of discussion.

A debate regarding the true extent of the political, cultural and social influence and the economic power of a large television broadcaster such as TV Globo is quite complex and wide-ranging; yet, despite its relevance, lies outside the scope of this research. This book will analyze the transformations of spectatorship of traditional fictional contents of broadcast television that, in turn, currently transcend the boundaries of their narrative format and also of their original support or device, taking into account the media ecosystem, the participatory experiences and screen convergence. Coined by Scolari (2009, 2019), the term "media ecosystem" designates the set of media, platforms and screens used in the circulation, distribution, interaction and exhibition of content in a unique or convergent way. It is a way of accounting semantically for the multiplicity of supports, genres and constitution of different possibilities of the image and interactivity with the public.

The main question asked here is how Brazilian television and its massive viewership, having as an outline its most traditional content, the telenovela, can or cannot transform itself and survive the arrival of digital platforms with segmented content. How does the Brazilian telenovela resist, negotiate and adhere to new platforms and transformations in spectatorship?

To answer this question, we must first define what this book understands by "television" in Brazil, what the importance of the telenovela is to that definition, and moreover (indeed mainly) why the transformations of this content and its viewers' spectatorship are vital in speculating on the future of television in Brazil. The hypothesis – based on an analysis of the telenovela's format, its influences, consumption, characteristics and transformations – is that, as long as the telenovela continues to be part of Brazilian culture, it will continue to be a mass product.

Many European authors use the terms paleotelevision and neotelevision (Eco, 1984; Casetti and Odin, 1998; Scolari, 2009) to describe the transition between an exclusively public European TV with more pedagogical and educational needs, to the private, commercial television that succeeded it in the mid-1970s, with another programming format and flow. European paleo-television had the philosophy of public service; neo-television had a capitalist perspective of profit. Scolari (2009, 2019) coined the term hyper-television, which would not be the new stage of the paleo/neo series but, instead, a particular configuration of the socio-technical network around the medium of television. In hypertelevision, viewer control is more significant. Beyond changing channels, the viewer may interact with the content in other media.

In analyzing US television, several authors (Williams, 1974; Fiske, 1987; Lotz, 2007; Mittel, 2015) use the term Network Era to describe the period between the 1950s and the mid-1980s in which three television networks dominated the country's audience. Williams (1974) argues that a typical television experience was the linear flow of programming as a whole to the detriment of separate programme consumption. In many ways, this period is very similar to that of hegemonic television in Brazil today, in which four broadcasters dominate the more significant part of the audience: TV Globo, Record, SBT and TV Band. However, television changes in each country according to its specific contexts, and Brazilian television possesses unique characteristics, as we shall see ahead. Lotz (2007) describes the way the Network Era of mass media in the US gave way to the Post Network Era and the fragmentation of the viewing audience in a wide variety of cable television stations specifically conceived for niche audiences. This period is also marked by changes at a material level even as content and services become less linked to the physical presence of the television set. This transformation also affected American network television content. Mittell (2015) states that expectations of how viewers watch television, how producers create stories, and how series are distributed have changed, leading to a new model of television that he calls complex TV. To him, television becomes complex as narratives are required to fulfil the expectations of a more attentive multiplatform audience, given that a programme may be viewed many times and content additionally pursued on various available screens.

We must rethink what it means to watch television, the group of practices associated with it and what exactly constitutes television content before analyzing if this medium is coming to an end.

Jérome Bourdon analyzes discourses of the end of the medium and observes that the end of television started to be discussed as desirable very early in history, with a remarkable core of common arguments across countries. Bourdon considers that "the discourse of the desirable end of television cannot be separated from the long perception, at least among cultural and political elites, of television as a 'bad medium' or as a 'bad object'" (2018). Although admitting that expectations regarding television have changed, Lotz (2007) ponders that the transformations in the forms of watching audiovisual content have not accelerated the end of television but, instead, revolutionized that medium. She refers to US television, but here, I propose to reflect on the process of Brazilian television.

Brazilian television broadcasting was inaugurated in the 1950s, following the transmission format of radio, offering only local programming and transmission. Although television stations are public concessions, television in Brazil was a private enterprise from its onset; business groups that commanded media corporations – newspapers and radio stations – ventured into the new medium. The first station, Tupi TV from São Paulo, owned by the powerful media mogul Assis Chateaubriand, sold the equivalent of one year's worth of advertising to four different companies to finance its investments. The beginning of Brazilian television was, therefore, the opposite of the same period in European television. As the business grew and revenue increased, Brazilian television broadcasters became production companies, in addition to exhibiting and distributing content. To this day, the most significant share of open television content is produced in studios that belong to major corporations. Competition and new transmission possibilities led the local television broadcasters to become national networks.

Public television stations emerged in Brazil only during the 1970s, but they never achieved the same audience numbers as private broadcasters. Currently, the majority of television stations in Brazil, as in other parts of the world, seek technological solutions for online consumption of their content, whether through the live transmission of streamed programming, or through the exhibition, with hybrid business models, of programmes belonging to their collection and original productions on digital platforms. The transmission itself of audiovisual content for television has also transformed, as did cable or satellite transmission, which until recently was the primary form of distribution of what was seen on television.

Nevertheless, Brazilian broadcast television currently bears many resemblances to economic strategies of the period described by European theorists as neotelevision. Brazilian broadcast television also possesses massive audiences, in what is described as the Network Era by American authors. However, this does not mean that the Brazilian public does not have access to services that would be characteristic of more recent periods of television in the US and Europe. Although comparable in many ways to the theories and descriptions of various authors about stages in television history in different periods and countries, the Brazilian model is unique, and this includes its most traditional content: telenovelas.

According to Martín-Barbero (1997, 2004), the telenovela is not merely a well-established Brazilian television genre; it is the most important and longest-running programme format in Brazilian television, upheld by rules of melodrama and constitute nothing less than the Latin American narrative matrix. Martín-Barbero argues that it is a mistake to think that television would be a subject of communication rather than of culture, given that signs of Latin American cultural identity are recognizable in telenovela melodrama, recovering folk memory through industrial content. To him, no other genre has been as successful in captivating the region's audiences. Hamburger (2011) observes that "the Brazilian telenovela challenges polarization between high and low culture, classical and popular culture, modernism and mass culture." In Brazil, the notions of broadcast television and telenovela are mixed and are interpreted to be the same.

Forty-one telenovelas were broadcast on Brazilian television in 2016, 21 of them new on TV Globo, TV Record, SBT, TV Band and TV Brasil. There were 16 telenovela time slots per day. If we also take into consideration the Viva cable network channel linked to the Globo group, which reruns telenovelas, there were 5,431 exhibition hours in 2016. Of the 84.7 million people who watched the genre that year, 64.7 million viewers (approximately 80% of the viewing public) watched TV Globo telenovelas.

As a writer and director, whether in independent film productions or work for television broadcasters, I participated directly in the elaboration and production of several audiovisual works – feature-films, television series and programmes, and web series. Among those, I wrote four telenovelas for TV Globo: two seasons of *Malhação/Young Hearts* (2012 and 2014 – the first season was a nominee for the Digital Emmy awards in 2013, and the second for the International Emmy Kids awards in 2016 and in 2017), the teleno-vela *Totalmente Demais/Total Dreamer* (2016 – International Emmy nomi-nee for best telenovela in 2017), and, finally, the telenovela *Bom Sucesso/A Life Worth Living* (2019). Through them, I also took part in several initia-tives for extending this content to other screens. Therefore, this is a hybrid work that mixes empirical experience and market sources with theoretical sources. Monaco (2010) ponders that some academics examine their multi-ple positions as fans (of series) and researchers (of series as well as of their fan groups) as vulnerable and hide them from other scholars. She considers the advantages of being explicit about the process and how localized iden-tity and emotional recall inform research choices. Hills' research on fans (2002) also considers that academia usually rejects the idea of hybrid identi-ties – that may unite not only within the academy but also outside it. In my case, the hybrid identities are those of scriptwriter and audiovisual director and that of researcher. I have chosen to explore both identities because of my belief that exposing the processes in which I participated would also contribute to the research at hand. During the last few years, I have worked intermittently for TV Globo as a director, series, and telenovela and trans-media writer. But this does not mean that I participate, approve of or even have access to the company's strategic decisions regarding screen conver-gence or any other subject. I only participate in the creative and artistic con-tent of the works with which I am involved. In this research, suppositions regarding television's business model are based on market movements, data analysis and research, legal matters and government regulation in counter-point to the network's initiatives. The following chapters analyze some of the empirical experiences in which I have participated in addition to similar experiences in writing for daily episodic television.

The second chapter of this book discusses telenovelas in Brazil from a historical and cultural point of view, the transformations the narrative expe-rienced and which attributes that remain the same until the present. Vari-ous agents influenced the telenovela narrative throughout its history, with emphasis on melodrama, feuilleton and the spectator. Even after almost 70 years of existence, the telenovela continues to be the central narrative of a massive audience in Brazil. Telenovelas are also exported to several other countries and territories. The chapter additionally analyzes the main char-acteristics of telenovela storytelling: themes, format and language elements.

The third chapter seeks to demonstrate how a telenovela is developed, from the first idea to the structuring and production of episodes. This chap-ter deepens and details the writing process which is essential to the format and also to the chain of power of the telenovela. In this chapter, empirical practice and experiences have more weight than in the remainder of the book. The reason is the lack of theoretical material on narrative construction and the process of writing Brazilian telenovelas in contrast to vast material regarding the process of the narrative construction of series and film scripts.

The fourth chapter investigates the public's relationship with the telenovela from the perspective of what a viewer expects from daily drama and how narrative adapts itself to these expectations; the possible dialogue between this viewer and the producers and network executives; what this viewer's power effectively is; and how it has been increased (or not) by new technologies and interactive platforms.

The fifth chapter of this book deals with the telenovela's business model and the consumption of narrative, of its characters, of its world. This theme has various layers, both the telenovela's relationship to a romantic ethics of consumption, as well as the appropriation by the public of postures and attitudes portrayed in television and the subsequent re-signification of the narrative. Content may also be seen as a product shared and understood by fans as a gift. This chapter will also tackle the intertwining of consumer products and the telenovela narrative as part of a business model.

The sixth chapter, made up of four case studies, explores the viewers' relationship to the telenovela and how new technologies have transformed this relationship and created a new balance of interaction. They are experiences related to the new spectatorship, to transmedia content, to the phenomenon of screen convergence, and slippage of content between distribution and exhibition platforms. Initially, we analyze the first experience of transmedia narrative content produced for a 9PM telenovela, the main product of Brazilian network television today; then an experiment with fanfic and telenovela; next, a spin-off of a telenovela for the internet; and, finally, the sliding of a 9PM telenovela to Globoplay's digital platform. The existence of a Brazilian-style convergence of media and the development of the transmedia experience in Brazil are under consideration here.

Finally, the conclusion of this book re-examines the main questions raised and discussed across its chapters: can the telenovela be the key to the survival of Brazilian television industry, and how can this narrative survive the sliding of content to other screens and media, the arrival of digital and interactive platforms and the transformations in spectatorship?

Bibliography

Barbosa, M. (2010) Imaginação Televisual e os primórdios da TV no Brasil. In: Sacramento, I., Roxo, M. and Goulart Ribeiro, A. (eds.) *História da Televisão no Brasil*. Rio de Janeiro: Editora Contexto. pp. 15–36.

Bourdon, J. (2018) Is the End of Television Coming to an End? *VIEW Journal of European Television History and Culture*, Vol. 7 (13), p. 3.

Casetti, F. and Odin, R. (1998) Da paleo à neotelevisão: uma abordagem semiopragmática. *Ciberlegenda*, Vol. 27.

Eco, H. (1984) Tevê: a transparência perdida. In: *Viagem na irrealidade cotidiana*. Rio de Janeiro: Nova Fronteira.

Fiske, J. (1987) *Television culture*. London; New York: Methuen.

Hamburger, E. (2011) Telenovelas e interpretações do Brasil. *Lua Nova*, pp. 64–74.

Hills, M. (2002) *Fan cultures*. London: Routledge.

Lotz, A. (2007) *The television will be revolutionized*. New York: New York University Press.

Martín-Barbero, J. (1997) *Dos meios às mediações*. Rio de Janeiro: Ed. UFRJ. p. 174.

Martín-Barbero, J. and Rey, G. (2004) *Os exercícios do ver: hegemonia audiovisual e ficção televisiva*. São Paulo: Senac.

McLuhan, M. and McLuhan, E. (1988) *Laws of media: The new science*. Toronto: University of Toronto Press.

Mittell, J. (2015) *Complex TV: The poetics of contemporary television storytelling*. New York: New York University Press.

Monaco, J. (2010) Memory work, autoethnography and the construction of a fan-ethnography. *Participations, Journal of Audience & Reception Studies*, Vol. 7 (1).

Muanis, F. (2018) *A imagem televisiva- autorreferência, temporalidade, imersão*. Curitiba: Appris Editora.

Scolari, C. (2009) Ecología de la hipertelevisión: complejidad narrativa, simulación y transmedialidad en la television contemporánea. In: Squirra, S. and Fechine, Y. (eds.) *Televisão digital: desafios para a comunicação*. Porto Alegre: Sulina.

Scolari, C. (2019) Narrativas Transmídia: consumidores implícitos, mundos narrativos e branding na produção de mídia contemporânea. *Parágrafo*, Vol. 3 (1), pp. 7–20. [online] Available at: http://revistaseletronicas.fiamfaam.br/index.php/recicofi/article/view/291. Accessed: 3 February 2019.

TV Globo. (2017) *Globo celebra alcance de mais de 100 milhões de pessoas por dia*. [online] Available at: https://redeglobo.globo.com/novidades/noticia/globo-celebra-alcance-de-mais-de-100-milhoes-de-pessoas-por-dia.ghtml. Accessed: 6 October 2018.

Williams, R. (1974) *Television*. London: Fontana.

Wolton, D. (1996) *Elogio do grande público – uma teoria crítica da televisão*. São Paulo: Ática.

2 Telenovela
History, transformations and resilience

The feuilleton, marked by the fictional and entertainment tone, appeared in France in the 1830s. According to Bragança (2007), with the consolidation of the bourgeoisie, the purpose was to expand the newspaper market. The feuilleton was located in the footer of the newspaper's front page and was a space destined to varieties. Soon the commercial potential that space would have in the very structuring of the newspaper and the relationship with the daily news became obvious. It is also from that time the first advertisements were inserted in the pages of newspapers, emphasizing the beginning of a market press. This relationship would be fundamental in the constitution of the cultural industry in the 20th century, present in a mass culture produced, for example, by the North American soap opera, which emerged in the 1930s, and the radionovela, which also bears a strong connection between the public and content and later the telenovela. Until this day, Brazilian telenovelas, in so many ways descendants of the feuilleton, still air between newsreels and advertisements in the linear programming flow of corporate television.

In Latin America, the feuilleton absorbed the French matrix but addressed issues of the local culture. Silva (2010) analyzes the repercussions in Brazil, where the feuilleton novel represented literary modernity through authors such as José de Alencar and the later migration of the genre to radionovelas and then telenovelas. The romantic sensibility – in its imbrications with the feuilleton and melodrama – remained present. According to Silva (2010), in the 19th and 20th centuries, melodrama underwent several transformations in search of ways to captivate various audiences, including those who began to attend the theatre in the wake of social transformations. Melodrama can adapt to the moment and the language and modernize itself to the taste of the times, always with a commercial bias. The serial aspect of the feuilleton along with fictional tools such as cliffhangers and the purpose of entertaining are still present in telenovelas. So are the characteristics of the melodrama narrative, such as its archetypes, heightened and polarized actions and dialogues. Brook (1976) observes that melodrama is radically democratic, striving to make its representations clear and legible to everyone. So do the telenovelas.

Even after almost 70 years of existence the telenovela continues to be the central narrative of a massive audience in Brazil, attracting millions of viewers daily. Telenovelas are also exported to several other countries. Hamburger notes that: "The export of Brazilian telenovelas demonstrated the possibility of reversing transnational flows of information and culture" (2011). The telenovela has been influenced in several ways that have made its narrative, language and format genuinely Brazilian. Hamburger (2011) ponders that Brazilian professionals sought to distinguish their work from other Latin American productions. According to the author, whereas Mexican telenovelas are more melodramatic, Brazilian telenovelas have a more naturalistic trait, investing in shooting on location and making use of colloquial language. Brazil has imported texts from Cuba, like many other countries, for radionovelas and afterwards for the first telenovelas. Nevertheless, because they were considered excessively dramatic, productions were only transmitted after undergoing adaptations.

Like the Brazilian telenovelas, the US soap opera also has a melodramatic matrix and uses narrative tools with origins in the feuilleton, such as cliffhangers. In addition to these characteristics, soap operas also have in common with Brazilian telenovelas – and most Latin American telenovelas – the fact that they have had a business model since the beginning supported by advertising, mainly of products associated with the home and the female audience. In Brazil, advertising of brands and products also played an important role, leveraging and sustaining daily TV drama. Almeida (2002) investigates the association of telenovelas with female audiences, which would be an unfolding of older associations between the feminine and specific cultural productions, such as the feuilleton and melodrama, demonstrating that there is great commercial interest in this symbolic construction.

Telenovelas in the early years in Brazil were supervised by advertising agencies, and authors like Ivani Ribeiro and Walter George Durst worked for companies such as Gessy Lever or Colgate-Palmolive. In the book *Autores* (Fiuza and Ribeiro, 2008), Benedito Ruy Barbosa recalls: "I was the one who gave orders because I worked for Colgate, the company that sponsored everything. It was like this until [TV] Globo began to take care of productions, hiring authors." Ruy Barbosa was one of the many authors hired by Globo. However, Brazilian telenovelas also have many differences from soap operas. The main one is that they have a beginning, middle and end, while soap operas have no final prediction. *Days of Our Lives*, from the American broadcaster NBC, for example, is one of the most extensive works of television dramaturgy in the world: it began airing on NBC in 1965 and ended in 2013. The dramaturgy revolved primarily around the saga of two families, the Hortons and the Bradys. One of the actresses, Suzanne Rogers, celebrated 40 years of participation in this soap opera in 2013. In addition to this distinction, in Brazil, telenovelas are aired in prime time, while soap operas occupy a less prestigious slot in the afternoon. Telenovelas in Brazil can be current, period or even set in an imaginary universe, whereas soap operas are usually set in the present day.

Silva (2010) notes that the American stories at first adopted the style of the 19th-century English domestic novel, presenting middle-class family conflicts from a female perspective, while productions for radio and later for television in Brazil were adapted to the interests of Latin American women.

Over the years in the history of the Brazilian telenovela, the primary influence that singularizes it from others is, in fact, the audience that watches and engages with the content, giving it new interpretations. It is the viewer who ultimately stirs changes in the telenovela, a product of commercial television that depends on the massive audience because of its business model. Therefore, the interpretation by the broadcaster of what the viewer "wants" directly influences the content and programming. Authors need to anticipate themes that will be dear to the audience in the near future to develop a telenovela that will interest, provoke and interact with this audience.

In this chapter, the main characteristics of the telenovela – themes, narrative, format and language elements – will be addressed. It will discuss influencing agents that have acted throughout telenovela history, with emphasis on the viewer. The purpose is to analyze the transformations, attributes and elements that remain in this narrative through time.

2.1 Telenovela and society

Sua Vida me Pertence/Your Life Belongs to Me is considered the first telenovela of Brazilian television. It premiered on 21 December 1951, a little over a year after the inauguration of TV Tupi São Paulo and months after the inauguration of TV Tupi Rio de Janeiro by media mogul Assis Chateaubriand. Walter Forster wrote, produced, directed and also acted in the telenovela alongside Lia de Aguiar and Vida Alves. Together, Forster and Vida starred in the first kiss in Brazilian TV history.

> Walter Forster said to me once: "A telenovela. I'm going to release a telenovela on television." The female audience is growing day after day, and women love romance, love triangles. Suspense. And each chapter has to end up on top, to hook up onto the next chapter. *Oh, and I'm going to kiss you. It's going to be great!*
>
> (Alves, 2008)

There were many obstacles that Walter Forster faced in approving the kissing scene. Ricco and Vannucci comment: "The subject was taboo not only for the public but also among artists" (2017). Walter Foster had to assure Vida Alves' husband, an Italian engineer, that there would be no rehearsal and that the scene movements would be rehearsed technically. According to Vida Alves (2008), the chief executive of TV Tupi himself, Costalima, was against the scene, but one of the main arguments of the author, producer, director and actor Walter Forster was that in the United States this was already happening in films. Costalima, however, pondered that not only was the Brazilian public different from the American audience, but also that the programme would air in Brazilian homes, in the living room, and not in a dark cinema. After involving the entire board of TV Tupi, the kiss was approved.

Despite being chaste, according to Alencar, the kiss "generated protests of all kinds against the immorality that threatened the homes of the country" (2002). Vida Alves recalls the repercussion of the kiss. "Some people were scandalised, some kept quiet about it. And some were scared. Something important had happened; everyone knew about it" (Alves, 2008). As the telenovelas at the time were aired live, there is no documentation of this historical fact; there are only photos alluding to what happened that document moments before and after the kiss.

This episode illustrates how much telenovelas have changed in the last decades and how society influences these changes. A kiss would hardly scandalize the viewer of a telenovela nowadays. Depending on the time slot, in addition to several kisses, scenes with allusions to sexual acts may also appear. According to Eneida Nogueira, research director of TV Globo until 2017, "It is not people who watch television. It is television that watches society. What works on television is what society wants to see" (Svartman and Nogueira, 2018). According to her, a television author has to have the ability and sensitivity to glimpse in which direction the *ethos* of society is going: "The great secret of television is that you understand what is latent in society. Not something that everyone already knows, because that does not cause any admiration, but it also cannot be something that is too far away; otherwise, people do not recognise it" (ibid.). Following this reasoning, the kiss seen by families in their homes in 1951 also portrayed a tendency to greater permissiveness of society, a latency perceived by Walter Foster. Eneida Nogueira also remarks that the definition of what is a popular heroine has changed, especially in the last ten years. For the protagonist of today, romance cannot be the only goal; she also needs to have personal achievements, for example, in her profession. "People want to be inspired," says Nogueira (ibid.). The heroine in the telenovela needs to follow the transformations, struggles and accomplishments of the women of her time.

For telenovela author Aguinaldo Silva in *Autores* (Fiuza and Ribeiro, 2008): "The only characteristic of the young lady that remains is that when she falls in love, she is faithful to her passion." One of the main tools for broadcasters to try to understand what the public wants are focus groups. In TV Globo, this research usually happens a month after the telenovela start date. The rejection or acceptance of a romantic pair by the audience can define the romantic trajectory of the heroine. In the case of the telenovela *Caminho das Índias/India: A Love Story* (2009), by Glória Perez, the lead female role of Maya Meetha, played by Juliana Paes, was to have as a romantic partner Bahuan Sundrani, played by Márcio Garcia. However, after the first month, rejection of the couple caused the author to change the original story and the heroine then fell in love with Raj Ananda, played by Rodrigo Lombardi.

Eneida Nogueira notes that when television delivers something that goes against the viewers' principles, he or she loses interest. "TV does not impose. What goes well is what society wants" (Svartman and Nogueira, 2018). There is a big difference between what the public "loves to hate" and characters or situations that the public simply is repelled by and then stops watching. It is necessary to understand this viewer's limits and moral and ethical boundaries, without necessarily judging, but it is possible to foster dialogue through television dramaturgy. When the telenovela is perceived as reactive to what the viewer believes, imposing a way of thinking, the viewer rejects it. Therefore, it is necessary to understand that the Brazilian society of 1951 sought to see a kiss on television; the proof of this is that the telenovela was an audience success at the time, even though a large portion of the audience criticized the show.

In an interview for *Autores* (Fiuza and Ribeiro, 2008), Benedito Ruy Barbosa considers that, when addressing specific topics, it is necessary to reflect on the differences present in the various Brazilian regions:

> If every audience came from the well-to-do Copacabana, Leblon, or Paulista Avenue, you would be freer to address specific topics. But you also have to think about the inland viewers from Mato Grosso do Sul, Goiânia, Roraima, Acre. They are people who have a different, more conservative and moralistic background, who deal with their children differently.
>
> (Fiuza and Ribeiro, 2008)

Eneida Nogueira recognizes that, unlike journalism, which reports what happens in the world, the telenovela allows viewers to observe what happens inside their homes, around them, in their lives: "Fiction is the moment when the person re-signifies the things he or she thinks, the things he or she knows. It is like using that to reevaluate what you think, or reaffirm, or challenge, or put yourself in the shoes of others" (Svartman and Nogueira, 2018).

There are several examples of telenovelas that were rejected by the audience. What these telenovelas have in common is precisely not being in tune with the practices and principles of society at the time.

In *Dono do Mundo/The Owner of the World*, a telenovela written by Gilberto Braga that premiered in January 1992, the rejection towards Márcia, played by Malu Mader, made the author reformulate the whole plot. Gilberto Braga is a telenovela author known for discussing the morals and ethics of society. Works such as *Vale Tudo/Anything Goes* (1988) and *Pátria Minha/ My Homeland* (1994) compose a trilogy with *Dono do Mundo/The Owner of the World*. In Fiuza and Ribeiro (2008), Gilberto Braga tells us that the focus group of the telenovela revealed that the audience considered that the heroine was frivolous for allowing herself to be seduced by the protagonist with dubious morals. The first chapters of the telenovela revolved around a bet. The protagonist, Felipe Barreto, played by Antônio Fagundes, had made a bet that he could take the virginity of the young lady, who was engaged to one of his employees, before her future husband. He wins the bet, and the betrayed groom kills himself. Gilberto Braga comments that he considered the story very strong, although wrong as a television telenovela because it was cruel and therefore disturbing for the viewer. In an interview for the newspaper *Estado de S. Paulo*, Malu Mader (Fernando, 2014) attributed the rejection to the lack of consonance with the conservative and prejudiced society of the time. She said it was difficult to follow a different path for the character from what she and the team had planned. At the request of the television station, the author Silvio de Abreu interfered in the work, changing the profiles of the main characters, thus gradually recovering the ratings.

A case of a more recent and less obvious rejection of a telenovela was that of *Babilônia/Babylon* (2015) by the same author, Gilberto Braga. This time Braga co-authored the work with João Ximenes Braga and Ricardo Linhares. It is a less evident example because the low ratings, the primary thermometer of public disapproval, were due to more than one factor. Many critics attribute the rejection to the very lack of strength of the narrative – the duel between two villains played by the stars of the station, Glória Pires and Adriana Esteves. Ricco and Vannucci (2017) observe that at the time, critics blamed the excess of urban violence and proximity to reality as factors that put the audience off. Many preferred to switch over to the biblical plots of the competitor (Record TV).

According to a national survey (IBGE, 2010), 91% of the population in Brazil declares to have a religion. The biblical telenovelas of Record

Television Network address such values head-on. Record TV is owned by the Universal Church of the Kingdom of God, founded by Bishop Edir Macedo. It knew how to identify this opportunity and to offer well-made biblical telenovelas, with well-known actors, full of special effects that give credibility to these narratives. According to Ricco and Vannucci (2017), the language adopted in *Os Dez Mandamentos/The Ten Commandments* (2015–2016), without complicated dialogue or formal wording, is also an explanation for the high acceptance of the public. They further observe that the viewers pointed to the lack of violence and the themes inherent to every human being as the central values of this production. In an interview with Ricco and Vannucci (2017), the telenovela author Vivian de Oliveira comments that the dream of freedom of a people, faith and hope was all the public wanted. The author and the station perceived, at the time, a latency in society: talking about faith.

One of the authors of *Babilônia/Babylon* (2015), Ricardo Linhares, in an interview with the website Ego, recognizes that the kiss between the couple (the acclaimed actresses Fernanda Montenegro and Nathália Thimberg) in the first chapter drove away a part of the audience. He states that he does not regret the scene and comments: "There is no reason for such a stir. This reveals the prejudice and hypocrisy of Brazilian viewers" (Dezan, 2015). Regardless of whether Brazilian society is prejudiced, hypocritical or not, the episode also reveals the lack of understanding between the author and his audience.

In commercial television, low ratings can make a show go off the air. In the case of a telenovela, it is necessary to take into account the considerable investment that this product demands: hiring actors, crew, set construction. The best alternative, therefore, is to modify the work so that it pleases the audience again. The authors made several modifications to *Babilônia/ Babylon*. They diminished affection and explicit caresses among the characters mentioned earlier. Other characters also underwent changes: Carlos Alberto, played by Marcos Pasquim, was initially gay but eventually was written to fall in love with a woman; Alice, a character performed by Sophie Charlotte that would be a prostitute, turned into a romantic young lady. Ricardo Linhares ends the interview by correctly asserting: "It is necessary to always dare. The writer cannot settle for any less" (Dezan, 2015). It is the pursuit of the Brazilian telenovela to discuss themes pertinent to society, to be relevant, and to raise issues that are often not initially perceived, but that are part of the subjective dimension of Brazilian society. An author who seeks only to please the public and writes under self-censorship will hardly be able to offer the viewer an instigating story. The difficulty lies in the tenuous boundary between what promotes discussion – and therefore, interest – and what causes rejection. The author needs to have the sensitivity and inspiration to write in an unstable situation. Aguinaldo Silva ponders:

> When people approach me on the street, I talk to them. I am in permanent contact with the public. I'm the first to notice the signs that something's wrong. I have an immediate response from people. I don't wait for the outcome of the focus group to know what they are thinking about my telenovela.
>
> (Fiuza and Ribeiro, 2008)

Several kisses in homosexual relationships have recently been aired on Brazilian television, but usually a network of affection is created around the characters so that viewers, as prejudiced as they may be, desire them to be happy. Ricco and Vannucci (2017) write about the first gay kiss in a Brazilian telenovela in the last chapter of *Amor à Vida/Trail of Lies* (2014), by Walcyr Carrasco. During the telenovela, under pressure from the audience, the author transformed Felix, a character originally planned to be a villain who was capable of throwing his newborn niece into a dumpster, into a regenerated hero. Felix and his boyfriend Niko ended up together, happily forming a family and kissing. In an interview for Ricco and Vannucci's book, Walcyr Carrasco states that it was a significant step for individual freedoms and human rights. Nogueira (Svartman and Nogueira, 2018) considers that the relationship of the public with homosexuality in telenovelas has gradually changed. The relationship that viewers have with the characters is the same as they have with their children, friends and families – people they have affection for. Nogueira (ibid.) observes that, in recent years, gay women have begun to reveal themselves in focus groups without any embarrassment or friction with other participants. In the case of *Totalmente Demais/Total Dreamer* (2016), a telenovela I wrote with Paulo Halm, we tried to follow this strategy of affection with Max, played by Pablo Sanábio. That telenovela's focus group revealed that the public adored Max, and with this information, we increased the story of this character, creating relationships for him as a gay man. Felipe Cabral, a collaborator and LGBTQI activist, who wrote Max´s scenes, also suggested a sequence that would show how prejudice could make this beloved character suffer. The scene of the homophobic attack against Max and his boyfriend was a chapter cliffhanger. The effect of the scene on the audience was encouraging, with the public support of actors, non-governmental organizations (NGOs), the writing team and crew on social media. At the time the telenovela aired, homophobia was not a crime. Only in 2019 did the Supreme Court rule that discrimination against sexual orientation was illegal in Brazil.

To Fiske (1987), the ultimate power of a work's message lies in the viewer's reading of it, rather than in the producers' ambition or proposals. The relations between the texts take place in two dimensions – horizontal and vertical. The horizontal dimension exists between primary texts such as, for example, two different telenovelas. The vertical intertextual relationship consists in the relationship of a primary text to secondary texts that use the former as reference: advertising, criticism, promotional content. The vertical intertextual relationship may also extend to tertiary texts. These occur at the level of the spectator and his social relations. The researcher explains the public's relation to television, pondering that primary, secondary and tertiary discourses, from the perspective of the viewers' life trajectories, are associated with the content and favour the creation of this social link. A spectator relates the television text to his or her lived experiences, to reading their intertexts within their own historical and social contexts. Consequently, each reading is unique, and every viewer actively elaborates on these connections. The more subsidies the work offers for this phenomenon to occur, the higher the connection of the public with the work. A telenovela has the challenge and vocation of engaging with a massive audience, creating identification with plots, practices and characters in order to bring about reflection on everyday topics, in addition to offering information, entertainment, fantasy, decompression and inspiration. For Hamburger (2011), the telenovela becomes a privileged stage for the problematization of interpretations in Brazil because it can make daily chronicles based on conflicts of gender, generation, class and region. In 2012, given the repercussion of the telenovela *Avenida Brasil/Brazil Avenue*, Maria Immacolata Vassallo de Lopes, coordinator of the Telenovela Studies Center of the University of São Paulo (USP), observed: "The telenovela, in Brazil, more than entertainment, is the narrative of the nation. João Emanuel captured the spirit we live today" (Britto and Bravo, 2012).

Despite the approximation and harmony of telenovelas with society, it is necessary to have a critical look at the real representativeness of daily life and Brazilian society by telenovelas. In his documentary *A Negação do Brasil/Denying Brasil* (2000), filmmaker and activist Joel Zito Araújo demonstrates how several works distort reality and how even in literary adaptations, the characters are "whitened." He mentions, for example, the telenovela *Escrava Isaura/Isaura* (1976), in which the main character, the daughter of a black woman and a white man, is played by Lucélia Santos, a white actress, misrepresenting the description of the character in the original work. Recently, in *Segundo Sol/A Second Chance* (2018), written by João Emanuel Carneiro and directed by Dennis Carvalho, the casting of a majority of white actors was questioned in a telenovela set in Bahia, one of the states with the highest number of people declaring themselves as being black or mixed-race in Brazil. The headline of the *Huffington Post Brasil* Online was explicit: "'A Second Chance': The white Bahia of the telenovela is quite different from the real Bahia, with 76% of blacks" (Terto, 2018). *Veja* magazine showed the repercussions even before the work went on air: "The next nine o'clock telenovela of TV Globo, *Segundo Sol/A Second Chance*, has not debuted yet, but already has provoked controversy on social networks" (Veja, 2018). Regardless, the audience of the telenovela, which dealt with paternity, sexual orientation and incest, among other themes, was significant, and the outcry over the casting of a majority of white actors was diluted over the months on the air. The issue of representativeness and diversity in characters and plots becomes pressing in contemporary times, in which identity struggles are increasingly present. Telenovelas need to follow society and continue addressing values, practices and transformations, activating the public's secondary and tertiary discourses in order to survive. Therefore, they need to offer the viewer more and more narratives, plots and actors that can create identification, dialogue and harmony with the present, to remain relevant. In this sense, they need to absorb diversity, representation and identity demands and struggles in their narratives.

2.2 Characteristics of the telenovela

The first Brazilian telenovela, *Sua Vida me Pertence/Your Life Belongs to Me*, by Walter Forster, provides us with some clues of what characteristics and language elements remain and those that have changed. The melodramatic matrix is the main similarity between the first Brazilian telenovela and the current ones. The first Brazilian telenovelas followed the format of the feuilleton, already incorporated by radionovelas, with elements used until today. For example, every chapter has a main story that ends in a scene of great revelation, left open for the viewer to desire the next chapter desperately: the cliffhanger. It is usually a critical scene of high emotional impact: a reconciliation, a punishment, a devastating revelation or action without a clear outcome – an accident, for example – intentionally cut off at the height of the action. According to Alencar, "The end of the narrative at the key moment, which receives the most appropriate name of a cliffhanger, is an art, the art of making the viewer wait" (2002). It is the cliffhanger that offers the viewer the feeling that the story is advancing. Cliffhangers are also widely used in current series, fostering the practice of binge-watching on digital platforms. Author Glória Perez tells how she began writing telenovelas and soon reasoned that it was necessary to work with the rules of the feuilleton of the last century, which meant "using the cliffhangers that created the expectation of the next chapter and privileging the sensational over coherence" (Fiúza and Ribeiro, 2008).

There are also the so-called false cliffhangers that do not lead to a significant transformation in the story – when the sequence of the scene the next day does not bring real advances or changes to the plot, although the author tries to deceive the viewer to make it look like it would. A famous example is in Gloria Perez's telenovela *Carmem* (1987), written for the now defunct Rede Manchete, based on Merimée's short story. At the end of the chapter, the main character fired six close-range shots at her husband. The scene ended at the height of the action. However, in the next chapter, the viewer realized that no shot had hit the husband, either because the character had bad aim, or because she did not want to really hit him. As the act did not have a direct consequence on the narrative, it is called a false cliffhanger. The author Carlos Lombardi, responsible for several telenovelas, usually pays homage to this false cliffhanger of Gloria Perez in his telenovelas in a parody tone. He repeats the scene, and as always, the six shots do not hit the target (*Bebê a Bordo/Baby On Board*, 1989, *Perigosas Peruas/Dangerous Dowagers*, 1992, both for TV Globo).

When writing telenovelas, the author may have to use false cliffhangers at the end of a chapter – possibly because he or she could not find a scene of high emotion, action, transformation or impact at that moment of the narrative. There are many reasons for this: the chapter is too long and, consequently, needs to end before the author initially planned; it is too early to reveal a big secret; it is a chapter of preparation for a big plot twist. In such cases, a classic false cliffhanger would be for the protagonist try to tell a secret, but arrives too late, or feels unwell and eventually faints, for example. False cliffhangers should preferably be accompanied by action and never happen only in dialogue. There are two reasons for this: first, the action can help exacerbate the effect of a formerly weak cliffhanger; second, pleonasm is part of telenovela language, and for a weak cliffhanger, it is worth reinforcing dialogue with action. A critical element of telenovela language is pleonasm or reiteration, associating actions with dialogue, restating not only what happened in previous chapters but also the actions of the characters. There is a specific spectatorship for a telenovela: the public may watch the show while doing household chores, talking to the family or using social networks.

An element of radionovelas that remains today is the narrator. The narrator may also help strengthen a cliffhanger through a comment. The narrator, currently, is also used in telenovelas to make a quick recap of the previous chapter just before the next chapter begins. The recap is necessary because the public is not necessarily assiduous. As a telenovela chapter always ends with a cliffhanger, which usually is also the first scene of the next chapter, the summary of the previous chapter – or even the story – can better the experience, especially for sporadic viewers.

A striking feature of the telenovelas, the verisimilitude with everyday life in dialogues, plots, themes, and especially in imperfect and more realistic characters, has not always been present. As Bahia notes,

> The success and consolidation of the telenovela as the most popular and profitable genre of television was focused on a change of language: gradual substitution of the theatrical and fanciful, by realistic language and themes of contemporary Brazilian daily life that is always updated by the fashion culture.

(Bahia, 2014)

A significant milestone of this transformation is the telenovela of the defunct TV Tupi *Beto Rockfeller*, 1968. To Carlos Lombardi, "Beto Rock-feller took us out of the nineteenth century and brought us to the twentieth century, or rather to the end of the 19th century. He came out of romanticism and entered realism and naturalism" (Fiúza and Ribeiro, 2008). It was the first successful telenovela with colloquial language, and 80% of the scenes were recorded outdoors, in real locations near the headquarters of Tupi TV, in São Paulo. Conceived by Cassiano Gabus Mendes, the telenovela was written by Bráulio Pedroso, a theatre graduate and a journalist for the news-paper *O Estado de S. Paulo*. At the time, the adaptations of foreign authors were still popular. Ricco and Vannucci (2017) reveal that Luis Gustavo, who played the leading male role, was allowed to improvise and this helped to create a naturalistic impression. According to the authors, "Recording a scene to be aired on the same day was something relatively common in *Beto Rockfeller* and required team strategy, much confidence from director Lima Duarte and the certainty that the tape would arrive on time" (Ricco and Vannucci, 2017). Even though this sounds like a producer's nightmare, it certainly brought freshness to the narrative, helping the telenovela feel up to date. The telenovela tells the story of Alberto, a young middle-class shoe salesman who lives with his parents and is dazzled by São Paulo's high society. He then comes up with the alias Beto Rockfeller and impersonates a third cousin of the US oil tycoon. Ricco and Vannucci (2017) ponder that by breaking with Hispanic original texts and seeking national elements and facts of our daily lives, Cassio Gabus Mendes and Bráulio Pedroso bet on something risky but fundamental to bring viewers together due to a greater identification with the story.

Director Daniel Filho ponders: "One thing, however, I am sure – television identifies intimately with the place and time where it is made" (2001). The contemporaneity of *Beto Rockfeller*, which can be trans-lated into verisimilitude with everyday life, in the approach of issues in tune with the society of the time and colloquial dialogue, is still a trait of current telenovelas – especially prime time ones, which are rarely historical pieces. According to Alencar (2002), it was also in *Beto Rockfeller* that the first merchandising appeared. As the main character drank whiskey, the merchandising was precisely the hangover remedy "Engov."

In his memoir, media mogul José Bonifácio de Oliveira Sobrinho (2011), "Boni," reports that one of the great obstacles he faced to reformulate TV Globo's telenovelas, making them more contemporary, was the supremacy of Cuban author Gloria Magadan. She began her career on the radio in Havana and then worked on several Latin American telenovelas through the advertising department of Colgate-Palmolive, a company that was one of the main sponsors of the first Brazilian telenovelas. In 1965, TV Globo hired her to direct the newly created telenovela department at the invitation of the then director-general of the station, Walter Clark. Gloria Magadan thought telenovelas should avoid reality, privileging romantic fancies and, in general, set in distant scenarios – just the opposite of what Boni wanted to do in TV Globo, influenced by the success of the telenovela *Beto Rockfeller* on another station, TV Tupi. Gloria Magadan, however, had a contract, signed by Walter Clark before Boni entered Globo, which gave her perennial powers to choose and supervise what would be the telenovelas produced by the station. Boni discovered a gap in the contract, which did not stipulate the number of chapters that each telenovela should have. According to Boni (Sobrinho, 2011), with the director Daniel Filho's support, he reduced the chapters of all her telenovelas until she left Globo, unable to tell the stories the way she wanted. Gloria Magadan thought they would have to hire her back, but Boni was already renewing the network's team of authors. Years after the event, Boni ponders about Gloria Magadan's contribution to the success of the format on the network: "Without Gloria's work, it would have been much more difficult to start from scratch" (Sobrinho, 2011).

The departure of Magadan and the end of her text supervision for all telenovelas made Globo approach contemporary Brazilian themes. Period telenovelas continue to be produced and exhibited, especially just before prime time, but in general, prime time telenovelas are contemporary.

Over the past decades in Brazil, telenovelas have undergone several transformations. While the first one had 15 chapters, aired on only two days of the week, today, a prime time telenovela will have no less than 150 chapters, airing Monday to Saturday. Thus, the risk of so-called *paunches* increases, which is when the plot of a telenovela does not seem to advance. Besides, a contemporary story will undoubtedly feature more than one narrative core. Concerning the narrative, the viewer is less and less patient, therefore scenes are shorter, and there is more than one story going on, even though chapters may be up to one hour long. This impatience of the viewer alters the dynamics of the telenovela. Carlos Lombardi, in an interview, considers:

"The chapter has changed in size, for two reasons: first, because it lasts longer on the air; second, because the narrative rhythm has changed. If you watch old telenovelas, you'll say, 'My God, what a slow thing! Was that the way it was?'" (Fiúza and Ribeiro, 2008).

The longer duration of the telenovela combined with shorter scenes has led to the increase of parallel plots: the different *nucleus*. According to Flávio de Campos, the "nucleus is the set of characters with a trait or a common circumstance" (2007). As a telenovela is an extensive work, which is written and produced while being exhibited, it becomes necessary to increase productivity. After all, if six one-hour chapters are shown per week, six one-hour chapters need to be produced per week. The strategy is several daily recording fronts, with different actors and sets. For this to be feasible, different stories must happen in diverse scenarios, with distinct actors, as much as possible. Alternating studio scenes with external scenes and in the scenographic city, it is conceivable to plan on several fronts. The material recorded daily goes to the post-production already pre-edited during the recordings. On the other hand, despite the various nuclei and secondary stories, the author cannot lose the focus of the main story that is responsible for most cliffhangers, if not all.

Mirian de Icaza Sánchez worked for 26 years at TV Globo Quality Central, analyzing research and products until 2016. Sánchez ponders that the main plot of a telenovela should be simple enough for the viewer to know how to tell it (Svartman and Sánchez, 2018). Daniel Filho notes that there is no rule for the success of a telenovela, but some aspects are fundamental. "We have to be simple in the way we present a story: it must be very clear and have few central characters" (Filho, 2001). He cites as an example *Rainha da Sucata* (1990), the first telenovela for prime time written by Silvio de Abreu, on which he was an actor and also an artistic supervisor. With an excess of characters and an unclear division between the genres of drama and comedy, the telenovela only got the desired audience after the author defined the main plot better. Silvio de Abreu also evaluated that the telenovela's audience increased when it became clear what the nuclei of humour and the drama was (Fiuza and Ribeiro, 2008).

Another vital element of current telenovelas, a result of the growth of the plot cores and nuclei, is the *repercussion*. When a significant plot twist occurs, it needs to resonate in the nuclei of the telenovela. Therefore, if a secret is revealed, it is necessary to build parallel action scenes in which the viewer can accompany the repercussions of this revelation. The audience expects to see the reaction of different characters to the same event. The importance of this element comes from the fact that the different nuclei often serve different portions of the public. The viewer wants to watch the character with whom she or he identifies most and see his or her reaction to that event. As professor Flávio de Campos ponders, "Movies and telenovelas often establish connections by empathy and identification with viewers. However, as the telenovela speaks to very disparate viewers, more than a film, the writer has to create different cores of characters through which different viewers can feel empathy or identification" (2007).

The *flashback* is an element present in the cinema and in the radionovela, but for the telenovela it became essential. The flashback of a scene has the function of reminding the viewer of a crucial moment of the plot. The first night the romantic couple spends together, for example, is a scene that the director shoots bearing in mind that it will be shown more than once. Telenovelas often have "sacred objects"; these are objects that refer to essential moments of the plot or feelings or relationships. For example, if the good guy gave the young lady a necklace, it becomes a sacred object. Sacred objects can trigger flashbacks. Then, when the young lady touches her necklace, the memory of their first night together is given by a flashback. Various elements of the audiovisual language are used to clarify that it is a flashback scene. They can have a different look from the rest of the scenes, the edges of the image may be blurred, or the dialogue may have a unique effect, and the transition usually is a fusion of images of the past and present: leaving the close-up of the character and going into flashback, for example. Many authors also use *fake flashbacks*. A fake flashback is a scene that has not appeared before in the telenovela, though it will have the same aesthetic that the director has chosen for the flashback scenes.

Nowadays, other transformations in the telenovela also stand out, especially with the recent influence of TV series, a format that has grown significantly abroad and in Brazil in recent years. Between 2009 and 2017, Ancine, a Brazilian audiovisual agency, reported 3,639 series produced in Brazil, including productions for broadcast and cable television. According to Martin (2014), in the book *Hard Men*, which chronicles the creative process of the American TV series since *The Sopranos* was released in January 1999, there are more complex stories and long-arc series with multifaceted characters. For the author, this is the beginning of a new golden age of series – which supposedly lasts to this day. Mittell (2015) also notes how the American series today have absorbed melodrama techniques and more extensive and dramatic narrative arcs. For him, this transformation is not necessarily the direct influence of American soap operas. According to the author, other serial works from different media, such as comic books, film franchises and 19th-century literature, have also influenced series in general and all, in turn, have their connections to melodrama. The taste for serials can be seen in the work of Brazilian telenovela authors. For example, João Emanuel Carneiro in 2015 gave a title to each chapter of his work, the telenovela *A Regra do Jogo/The Rule of the Game*. This is a format associated with TV series, especially the *procedural* ones, i.e. episodes with a new story. This influence is also present in the telenovelas I wrote. During the process, I realized the importance of plots that are solved on the same day, usually from secondary nuclei. This strategy makes the chapter exciting for a sporadic viewer. The telenovela has suffered various influences and transformations since 1951. However, the resilience of the format is due precisely to these changes that maintain massive audiences to this day that, in turn, make the product commercially viable. These changes are related to transformations in spectatorship, to the subjective dimension of Brazilian society, and, consequently, to the public that interprets, approves or rejects the stories of the telenovela. As long as the telenovela maintains this dialogue with the viewer, it will remain viable, whether on television or sliding to other media and screens, as will be discussed later in this book.

In the next chapter, we will analyze how a telenovela is developed, from the first idea to the structuring and production of the chapters. The process of script construction and production are the basis of the power chain of this product.

Bibliography

Alencar, M. (2002) *A Hollywood brasileira – panorama da telenovela no Brasil.* Rio de Janeiro: Senac. pp. 19, 62.

Almeida, H. (2002) Melodrama Comercial: Reflexões sobre a Feminilização da Telenovela. *Cadernos Pagu* (19).

Alves, V. (2008) *TV Tupi, uma linda história de amor.* São Paulo: Imprensa oficial do estado de São Paulo. pp. 114–115.

Bahia, L. (2014) *Telona e a Telinha: encontros e desencontros entre cinema e televisão no Brasil.* Ph.D. Thesis, Universidade Federal Fluminense–Instituto de Artes e Comunicação Social programa de pós-graduação em comunicação. p. 79.

Bragança, M. (2007) *Trópicos de lágrimas: um estudo sobre melodrama e América Latina a partir do cinema de cabaretera mexicano e da literatura de Manuel Puig.* Post-Graduate Program, Universidade Federal Fluminense, Letras.

Britto, T. and Bravo, Z. (2012) Especialistas explicam o fenômeno 'Avenida Brasil'. *O Globo.* [online] Available at: https://oglobo.globo.com/cultura/revista-da-tv/especialistas-explicam-fenomeno-avenida-brasil-6448625. Accessed: 12 December 2018.

Brooks, P. (1976) The 'Melodramatic' Imagination: Balzac, Henry James; 'Melodrama' and the Mode of Excess. *Modern Fiction Studies*, Vol. 23 (2).

Campos, F. (2007) *Roteiro de Cinema e televisão.* Rio de Janeiro: Zahar. p. 163.

Dezan, A. (2015) Ricardo Linhares avalia 'Babilônia' e não se arrepende por beijo gay. *Ego.* [online] Available at: http://ego.globo.com/televisao/noticia/2015/08/ricardo-linhares-avalia-babilonia-e-nao-se-arrepende-por-beijo-gay.html. Accessed: 25 November 2018.

Fernando, J. (2014) Malu Mader relembra reação conservadora do público em 'O Dono do Mundo'. *Om Estado de São Paulo*. [online] Available at: https://cultura. estadao.com.br/noticias/televisao,malu-mader-relembra-reacao-conservadora-do-publico-em-o-dono-do-mundo,1582226. Accessed: 25 November 2018.

Filho, D. (2001) *O circo eletrônico – fazendo TV no Brasil*. Rio de Janeiro: Zahar. pp. 11–69.

Fiske, J. (1987) *Television culture*. London; New York: Methuen.

Fiuza, S. and Ribeiro, A. (2008) *Autores: história da teledramaturgia*. São Paulo: Editora Globo. Vol. 1, pp. 36, 206, 241, 307, 308, 434.

Hamburger, E. (2011) Telenovelas e interpretações do Brasil. *Lua Nova*, pp. 64–74.

IBGE. (2010) *Censo 2010: número de católicos cai e aumenta o de evangélicos, espíritas e sem religião*. [online] Available at: https://censo2010.ibge.gov.br/noticias-censo?id=3&idnoticia=2170&view=noticia. Accessed: 22 July 2020.

Martin, B. (2014) *Homens difíceis – os bastidores do processo criativo de Breaking Bad, Família Soprano, Mad Men e outras séries revolucionárias*. São Paulo: Editora Aleph.

Mittell, J. (2015) *Complex TV: The poetics of contemporary television storytelling*. New York: New York University Press.

Ricco, F. and Vannucci, A. (2017) *Biografia da Televisão Brasileira*. São Paulo: Matrix. Vol. 1, pp. 23–274.

Silva, F. (2010) Melodrama, folhetim e telenovela anotações para um estudo comparativo. *Revista da Faculdade de Comunicação da Faap, São Paulo*, (15). [online] Available at: www.faap.br/revista_faap/revista_facom/facom_15/_flavio_porto. pdf. Accessed: 20 September 2010.

Sobrinho, J. (2011) *O livro do Boni*. Rio de Janeiro: Casa da Palavra.

Svartman, R. and Nogueira, E. (2018) Interview with Eneida Nogueira.

Svartman, R. and Sánchez, M. (2018) Interview with Mirian de Icaza Sánchez.

Terto, A. (2018) 'Segundo Sol': A Bahia branca da novela é bem diferente da Bahia real, com 76% de negros. *Huffpost Brasil*. [online] Available at: www.huffpostbrasil. com/2018/04/30/a-ausencia-de-atores-negros-em-segundo-sol-novela-da-globo-ambientada-na-bahia_a_23424010/. Accessed: 2 December 2018.

Veja. (2018) *Escalação de elenco de 'O Segundo Sol', nova novela das 9, é criticada*. [online] Available at: https://veja.abril.com.br/blog/virou-viral/segundo-sol-nova-novela-elenco-negros. Accessed: 2 December 2018.

3   The making of a telenovela

With an industrial production pace and commercial interests, the telenovela is a hybrid piece of work where narrative and creation play a significant role. Although it is a commercial endeavour, the Brazilian telenovela possesses distinctly authorial artistic features. The telenovela is written and produced while being exhibited, providing a dialogue of sorts with the audience. In order to write, the author must watch it and have a sense of its reception, either through research, social networks or his or her own perception. Thus, everyone depends on the author's rhythm. Whereas an author has less time to write a script, there is no time set aside in the production schedule for any author to be disputed – unless viewing numbers drop.

While describing the telenovela, the terms *quality* and *success* will be regarded as synonymous in this book. Since a telenovela is part of commercial television, success and quality both mean significant ratings. The Kantar Ibope Media polling group's National Television Panel measures audience samples in Brazil's 15 main cities, thus offering a market description of Brazilian television consumption. Their information serves as a pointer for planning a programming grid. In spite of the presence of specialized Brazilian television and telenovela critics, as well as theorists who push for greater complexity, sophistication or innovation in any given work, ultimately, it is the audience that keeps a telenovela on the air and therefore serves as a compass for this research. There is no precise formula for a telenovela's success, but there is clearly a construction process to that narrative.

Even though there is a rich theoretical base from the perspective of a script's construction and production process when it comes to film and television series, few authors concentrate on the development of telenovela scripts or production. Among those are Campos (2007) and Comparato (1996), who both include specific features of the telenovela in their writings concerning texts for the audiovisual market. Not many telenovela writers share their practical experience or the tools they use in writing daily drama. Thus, this chapter takes an empirical approach.

3.1 The search for subject matter as a starting point for narrative construction in the telenovela

To author Benedito Ruy Barbosa, "first and foremost, a telenovela must tell a big love story otherwise, not even men will like it" (Fiuza and Ribeiro, 2008). Mauro Alencar corroborates this: "From comedy to drama, the focus can be historical or social. But you've got to have a good love story" (2002). However, the love story may not be the starting point for writing a telenovela.

Doc Comparato (1996) lists six different fields where ideas may be found: the selected spark – which comes from memory; the verbalized idea – that we are told by someone; the read idea – that originates in a newspaper, magazine or leaflet; the transformed idea – that is born out of another work of fiction, a film, for instance; the proposed idea – commissioned to an author; and the found idea – through research or study. Three of these fields are more significant at the start of the process of working out a story for a telenovela. The first is that of the idea selected from memory. An intriguing character that motivates the author to write may be associated with his or her personal life. Author Benedito Ruy Barbosa ponders: "Evidently, most stories come from imagination. But the essential part comes from lived experience. Characters I came across, for example, in my travels" (Fiuza and Ribeiro, 2008). The author observes that the first part of his telenovela *O Rei do Gado/King of Cattle* (1996) was based on stories from his childhood. Memory and cultural repertoire may emerge in the narrative without the author's conscious awareness of them. An author has the freedom to write about whatever world or characters befits him or her; characters who are external to the author blend with his or her personal or secondary stories.

The second field highlighted is that of the read idea. Subjects that have the attention of the press or social networks may create a plot or a seductive universe around a character. Many authors have a small collection of clippings or digital archives of noteworthy news stories. Aguinaldo Silva, in Fiuza and Ribeiro (2008), for example, tells that the 1973 case of the missing boy Carlinhos inspired him to write *Senhora do Destino/Her Own Destiny* (2004). Alcides Nogueira, also in Fiuza and Ribeiro (2008), reports that he explores everything around him for ideas: books, films, plays, comic books, conversations in a dentist's waiting room or at the hairdresser. In a period telenovela, research combines with information that serves the same purpose, as in *Novo Mundo/New World* (2017), written by Thereza Falcão and Alessandro Marson. To Doc Comparato (1996), when research is involved, a new relevant field of ideas comes into play: found ideas.

Finally, the third field of stimulus for a telenovela would be the transformed idea. To write a telenovela, authors may draw inspiration from the classics and even from current contemporary works of fiction. In Fiuza and Ribeiro (2008), Gilberto Braga reveals that his inspiration for *Água Viva* (1980) came from *Annie*, an American musical. Alexandre Dumas' *The Count of Monte Cristo* (2003) has inspired many telenovelas, the most recent of which was *O Outro lado do Paraíso/The Other Side of Paradise* (2018) by Walcyr Carrasco. In the telenovela I wrote called *Malhação – Sonhos/Young Hearts – Dreams* (2015), the main story was inspired by William Shakespeare's *The Taming of the Shrew*, which we transported to the martial arts universe, mixing it with an art school scenario. These three fields of ideas are not separate; they function together. In another telenovela I wrote, *Totalmente Demais/Total Dreamer* (2016), for example, we began with a character living in a small town in the countryside of the state of Rio de Janeiro, who flees her pedophile stepfather and neglectful mother and goes to the capital. Her main challenge is to survive and, thus, reinvent herself. We were interested in the subject of pedophilia and had been gathering stories and accounts for a while. Writing about a survivor of one such case was, therefore, thought-provoking. The character's first romantic interest was a young man who she met in the streets and would help her to deal with the various misfortunes of her new reality. Not until later did we insert the Greek myth of Pygmalion into the story – the creator who falls in love with his creature. We decided to attach the character to the fashion world. Thus, it made sense that a casting agent could see himself discovering a promising new talent in the fashion business and "sculpting" her into a top model. Hence, the romantic triangle was made up of Jonatas, who calls himself "an entrepreneur" but actually sells candy on the streets; Eliza, the fugitive survivor; and Arthur, a casting agent in search of a fresh face. An editor-in-chief of a women's magazine, due to her love for the agent as well as her own interests, antagonized the protagonist. Yet the most significant antagonist to Eliza – played Marina Ruy Barbosa – was her pedophile stepfather and the paralyzing fear she felt of him.

Once the subject matter has been chosen and, based on it, the telenovela's premise is approved, the research process begins. To write a pre-synopsis – that is, the main story's principal thrust – without taking on secondary character clusters, the author may collect data and do his or her own research. But a researcher is essential for the final synopsis, with its secondary sets of characters and subplots, character descriptions and settings. A good researcher not only contributes data that is pertinent to the subject matter of the telenovela, but also brings up real characters who resemble those of the telenovela.

As authors, before production starts, when possible we seek to visit places that will be depicted in the plot and to speak to people personally, bringing to the narrative elements such as jargon, personal experiences and ideas for costumes and settings. It is also essential to be aware of the speech and rhetoric of real characters, never failing to observe differences among age groups. Frequently, before the telenovela starts, we try to ask our collaborators to visit the same places the fictional characters do and make lists of everything they consider interesting. During the elaboration of a telenovela, I usually cover the workplace walls with samples of dialogue, photographs, specific terms and pictures of the locations. When an author begins to write a telenovela, his or her workload and that of his or her team is exhausting, averaging 12 hours a day, at least six days a week, which doesn't allow him or her to go afield. During the telenovela, the researcher is the author's window onto the world and may bring accounts and stories from which to generate subplots or even obstacles in the main story, nourishing the author with information to sustain dozens of chapters. Research is also fundamental to the development of a plot that hinges on medical or legal facts.

3.2 Character composition and complexity in a telenovela

Martín-Barbero (1997) defines four main feelings at the central axis of melodrama: fear, enthusiasm, pain and laughter. Four types of situations allegedly correspond to them that are simultaneously personified by characters: the Traitor, the Righteous Man or Woman, the Victim and the Fool. When all of these come together, they produce a mixture of four genres: the romance, the epic, the tragedy and the comedy. "This structure reveals to us in melodrama a pretence of such *intensity* that it may only be achieved at the cost of *complexity*" (Martín-Barbero, 1997). It is possible to recognize melodramatic counterparts in a telenovela according to the clusters associated with the villain (the Traitor), the hero (the Righteous Man or Woman), the ingenue (the Victim) and the comic core (the Fool). Although this is the matrix, when conceiving a telenovela the creation of multi-layered characters is needed. A telenovela must keep track of societal transformations and, by extension, those of the viewing public.

Authors utilize many strategies to build and compose their characters. Aguinaldo Silva, for example, tells in the book *Autores* that Maria do Carmo, a character in *Senhora do Destino/Her Own Destiny* (2004), was exactly like his mother: "Her mannerisms, her words, the business of always being tremendously hungry, of being a matriarch and an authoritarian woman" (Fiuza and Ribeiro, 2008). In the same telenovela, his famous villainess, Nazaré Tedesco, was inspired by the cartoon *Tom & Jerry*, where everything Tom does turns back against him. "Nazaré was bad, but she did everything wrong. That's why she became funny" (ibid., p. 48). Benedito Ruy Barbosa reveals that his conversations with locals during the research for

Pantanal (1990) inspired him to create characters: "Suddenly, this returns to memory and you begin to create. That's how characters are born from these contacts. A lot is drawn from real life. Or from unreal life, who knows?" (ibid., vol. 1, p. 224). In Fiuza and Ribeiro (2008), for example, Giberto Braga reveals that it was the character of the drama critic played by George Sanders in the film *All About Eve* (1950) that inspired him to create the character of Renato Mendes in the telenovela *Celebridade/Celebrity* (2004). To Silva (2010), who analyzed Gilberto Braga's work, the premise of *All About Eve*, a young aspiring actress who presents herself as modest and naive in order to gain the confidence of the famous stage star whose place she will later take, is also present in the main plot of the telenovela. "Thus, in Gilberto Braga's appropriation of *All About Eve*, the great lady of the stage played by Bette Davis was transformed into the successful model and music business manager Maria Clara Diniz, played by Malu Mader" (Silva, 2010). To Silva, the character of Laura Prudente da Costa, played by Cláudia Abreu, has a profile even closer to that of her film matrix.

However, an author does not have full control over a character once the shooting has begun. A good relationship between actors, scriptwriters and directors builds a bridge between the text and the cast in the production of a telenovela. Aguinaldo Silva, in Fiuza and Ribeiro (2008) for example, remarks that Renata Sorrah contributed enormously to the success achieved by her character Nazaré Tedesco. On the other hand, in a panel discussion at the 2018 edition of São Paulo's International Biennial Book Fair, Walcyr Carrasco (Vicentini, 2018) revealed that he once sought revenge upon an actress who was improvising and changing the character. He wrote a throat condition into the plot so that the character would say nothing for two weeks. The actress punished by Walcyr Carrasco was Elizabeth Savalla, in *Chocolate com Pimenta/Pepper Chocolate* (2003), where she played villainess Jezebel. Many authors experience similar problems with actors who do not accept the fates of their characters, despite the director's mediation. On the one hand, the actor is on the air, and it is difficult to subtract him from the plot from one minute to the next, particularly if the viewing audience has taken a liking for him or her; on the other hand, the telenovela is an extensive work in which anything might happen: accidents, journeys, catastrophes. An excellent example of this was the earthquake caused by a volcano in a Janete Clair's *Anastácia, a Mulher sem Destino/Anastacia, a Woman with No Destiny* (1967), in which hundreds of characters died in a single chapter, thus making it possible for the author to change a problematic story that had begun with another author, Emiliano Queiroz, and raise audience ratings.

Eneida Nogueira, TV Globo's research director until 2017, recognized that throughout the years, viewers had increasingly come to expect more complex protagonists, with high ambitions and a need for personal and romantic fulfilment. For example, in *Totalmente Demais/Total Dreamer* (2016), Eliza's motivation was to help her family escape from her pedophile stepfather, hence her goal of making money; to this end, she enters a modelling contest. Throughout the telenovela, though, the heroine discovers that her mother had always known about her stepfather's moral faults and never thought of leaving him. Beyond this, because Eliza also discovers that she enjoys modelling, her motivation is transformed, and pursuing a career becomes her new goal. A good protagonist must have a clear goal or motivation, credibility to achieve his or her goal, and empathy with the viewer. A character's empathy bears a direct relation to the obstacles in his or her way.

John Truby (2008) suggests bringing ghosts to all the characters. The ghost may be a deceased relative. In *Totalmente Demais/Total Dreamer* (2016), Sofia, Lili's daughter, died in a severe car accident, and because of this Lili became extremely depressed, and her marriage and professional life were shaken. When the telenovela was extended by two weeks by TV Globo executives, we decided to bring the character of the daughter back to life, making up an explanation with fake flashbacks to show how she had escaped death. The perfect long-lost daughter came back as a sociopath, very melodramatic. I learned that having a ghost that can come back to life may help when having to extend a telenovela at short notice. That week the telenovela had some of its highest ratings. In 2016, each rating point was equivalent to 240,886 households or 684,202 people, considering the 15 main markets in the country. The average of the week was 35 points or an audience of approximately 24 million people per day.

Often, the ghost is not a character who has died. In the film *Thelma & Louise* (1991), an example used by Field (1995) to illustrate the need for elaborating a character's backstory, Louise's "ghost" is that she has been raped in the past. It is this fact that justifies her violent reaction when the character realizes that the same thing is going to happen to her friend, in the same city. In the telenovela *Totalmente Demais/Total Dreamer*'s 2015 season, the fact that Karina's mother, played by Isabela Santoni, died in childbirth, brings a ghost to the character, who mistakenly believes that her father rejects her because of this and loves only her older sister. An essential narrative tool is the ability to diagnose what the viewer knows before the character; what the character knows before the viewer; and what the character and viewer will find out together. The audience knew Karina blamed herself for her mother's death and therefore understood some of her actions. On the other hand, her father did not and this ignorance would prevent him from realizing why his daughter was so jealous of her sister, for example.

There is a straightforward rule to creating secondary characters or even antagonists: they must exacerbate the protagonist's most essential features, especially through their differences. For instance, if the protagonist is a young rebel, he will have conservative relatives; if the protagonist is conservative, he may fall in love with someone who holds libertarian ideas. Different characters such as these force the main one to continuously question and reaffirm his convictions throughout the plot and specially to generate conflict – the driving force behind a long, commercial narrative. To scriptwriter and researcher Flávio de Campos (2007), a character such as this may be called a *reverse character*, given that he highlights, through opposition, the hero's traits. Campos (2007) also points out the *ladder character*. The ladder character is the one who instigates the action of the main character with whom he is linked. The *ear-character* is the one who lends his attention – literally his ear – to the main character with whom he is connected. A secondary character that is the ideal complement possesses all three of these dimensions, under the premise that the most important of these is the *reverse character*. Even in the case of two or more protagonists, these might be one another's reverse characters, to use the term proposed by Campos (2007).

Janete Clair, one of the primary authors in the history of telenovelas, created a universe of characters by investing in reverse characters. The main character in *Eu Prometo/I Promise* (1983) is Lucas Cantomaia, played by Francisco Cuoco. He is an honest, conservative politician with immense credibility at the height of his political career, running for a seat in the Federal Senate. Because he needs to keep up an image of being a good father and husband, Janete Clair placed an irresistible photographer in his path. He falls in love and this puts his marriage at risk, setting off a crisis. Also, the author gave the politician's character a convicted brother of dubious mettle, one of Lucas' antagonists. This character questions the politician's work for an NGO that deals specifically with the reinsertion of former prisoners into the work market. Lucas' two assistants bring to the fore debates that lay bare the politician's moral dilemma, one forever offering positive advice and the other providing bad advice of dubious ethics. Additionally, Lucas had three daughters, among them an alcoholic and one who he soon discovers to be the child of his wife's extramarital affair. *Reverse characters* force protagonists to reaffirm their main characteristics through dialogue and action, as required by the medium of television. In *Bom Sucesso/A Life Worth Living* (2020), I started the synopsis with the story of Paloma, a humble seamstress, still young but a single mother of three who has a tough life but manages to be happy until she discovers she is going to die of a mysterious disease. She then regrets all the things she has not done in her life. The next character I wrote was her reverse. Alberto is a wealthy older man who did everything he wished during his life. He thought he was going to die of a mysterious illness but suddenly discovers he will actually live. Their tests had accidentally been mixed up. When she finds out she will not die, the seamstress wants to find the man that "will die in her place." Alberto and

Paloma meet and learn not only about their differences, but also what they have in common: a love for literature.

A character may help the protagonist to establish him or herself also by being a mirror. Therefore, I will add the *mirror-character* dimension and use Alice, from the same telenovela, as an example. Alice is the seamstress' daughter. Like her mother in her youth, she wants to go to university. Nevertheless, also like her mother, Alice gets pregnant while still in her teens. Seeing her daughter go through the same things she did also makes Paloma act and reveal herself as a character.

In a telenovela, the protagonist is not static and will change throughout the plot. The more *reverse characters* there are in the A-plot, or main plot, the more fluid this process will become, given that the protagonist will be questioned and stimulated to reaffirm his or her objectives and personality throughout the telenovela's trajectory. In *Totalmente Demais/Total Dreamer* (2016), Eliza, played by Marina Ruy Barbosa, starts as an immature, impulsive, frightened character who runs away from her pedophile stepfather after attacking him with a broken bottle. She steals money from the cash register, runs to the motorway, catches a ride in a lorry and winds up in the capital selling flowers on the streets. Towards the end, she is a rational, secure and successful top model.

Figure 3.1 Eliza, played by Marina Ruy Barbosa, selling flowers on the streets
Source: Globo | Renato Rocha Miranda.

Figure 3.2 Eliza, played by Marina Ruy Barbosa, in the final stages of the modelling contest. Source: Globo | João Miguel Junior.

All the main characters also undergo transformations. Eliza's "creator," the casting agent Arthur, a selfish and hedonistic character, learns about love and generosity. Unlike what occurs in a series or a film, the writer does not need to know how a telenovela character will change. A telenovela is a work in progress; it is written while being produced and broadcast.

3.3 The elaboration of a synopsis and the importance of plot twists in planning for works of long duration

A show bible for a series is a document that describes the story's world, tone, plots, scenarios, characters, the development of the season, narrative and visual references, very often the pilot script and the description of possible future seasons. The detailed bible of a series is similar to the synopsis of a telenovela. However, a season outline for a series will have an expected ending, whereas in a telenovela, multiple factors may influence the conclusion. Therefore, the end of the plot is not at all likely to be revealed in the synopsis; partly in order to maintain secrecy, partly because it is characteristic of the narrative to be open to transformation.

Audience rejection or approval, corroborated by qualitative research or low audience rates, may change the story of a telenovela. Another factor that influences the story is the *chemistry* between actors. Daniel Filho defines chemistry between actors thus: "When two actors engender great emotion in a scene, generating great power by the mere fact of appearing together, we say there is a certain chemistry between them" (2001). Also, the author's sensibility to changes in society during the months of a telenovela may alter the story. For example, Manuela Dias and Daniel Ortiz, telenovela authors that have had to pause their productions during the 2020 pandemic, incorporated COVID-19 into their narratives when the telenovelas came back. Finally, production problems caused by weather changes, illness and other factors may cause the author to make adjustments. During *Belíssima* (2006), for example, written by Silvio de Abreu, Glória Pires, who played Júlia Falcão, got hepatitis and had to take sick leave for three weeks. The author then invented that the character, traumatized by her grandmother's death, was admitted to a psychiatric clinic. While her romantic partner searched everywhere for Júlia, the author wrote more scenes for the secondary plot.

A telenovela synopsis will circulate among executives who may or may not green-light the telenovela. The introduction of the synopsis of the telenovela *O Clone/The Clone* (2001), by Glória Perez, an audience success in Brazil that has already been sold to more than 130 countries, made a synthesis of the topics that would be addressed before entering the story itself:

> The main set of our telenovela is the 21st century, the new millennium. It is often said that the twentieth century was the century of science and technology. These areas developed so rapidly that humanity cherished fantasies that in the year 2000 science and technology would be able to solve all problems and ward off the ghosts that have plagued man since his appearance on earth: material needs, disease, old age, death. There was a high expectation around that date as if it were the landmark of a new era for humanity. But as all optimism is reversed, the arrival of the year 2000 frightened many people: millennial prophecies indicated this date as the end of times. The apocalypse predicted by the Bible, by Nostradamus, by the Virgin of Fatima, by Rasputin, by the Mayan civilization, by the Muslims, at last, by all the prophets who have inhabited the earth. And here we are, coming to the new century, to the new millennium. The world is not over and we do not conquer, through science and technology, earthly paradise. The telenovela is about making this inventory: what has changed? Where are we heading?
>
> (Perez, 2001)

The use of epigraphs is also commonplace. Glória Perez, for example, begins the synopsis of *O Clone/The Clone* (2001) with a quote from Dostoevsky (1990): "If God does not exist, everything is permitted" (1990). In the telenovela *Bom Sucesso/A Life Worth Living* (2020), which dealt with, among other subjects, second chances in life, we chose a quote from Oscar Wilde: "To live is the rarest thing in the world. Most people exist, that is all" (2003). The synopsis then will describe the A-plot. Often a plot summary is presented in a few lines, prior to further development, in order to guide the reader.

A telenovela synopsis should have good turning points and narrative plot twists. This term is widely used by theorists and researchers to designate the moment in which the plot is entirely modified by the characters' actions, through an external event or a conflict that will reach its peak. A telenovela calls for at least three major turning points or plot twists along the way. This major turning point is developed into lower (or secondary) turning points and as weeks go by produces good cliffhangers per chapter.

To the viewer, these plot twists, turning points and cliffhangers translate into short, medium and long-term expectations. The paradigm would be one of several arrivals, which is to say, the author writes a chapter in order to arrive at the cliffhanger; the cliffhangers to arrive at the turning point, or revelation of the week; and the turning points of the week to arrive at the next big plot twist. A telenovela usually has more turning points, revelations and significant events in the beginning, the middle and at the end. Unless turning points come up in the plot, the story will seem static.

The Count of Monte Cristo, Alexandre Dumas' 1844 novel, has inspired many telenovelas, like *O Outro lado do Paraíso/The Other Side of Paradise* (2018), by Walcyr Carrasco, and *Flor do Caribe/Caribbean Flower* (2013), by Walther Negrão, because the essential plot twists are already present in the original narrative:

1   Edmond Dantès is falsely accused of treason on the day of his wedding to Mercédès and, after a troubled trial, is incarcerated on an island.
2   Edmond meets and befriends Abbé Faria who tells him of a hidden treasure.
3   Faria dies Edmond escapes and finds the location of the treasure. Edmond plans to reward those who tired to help him and punish those who hurt him.
4   Years later, Edmond spins a vast web of intrigue to gain revenge on his enemies, going under several aliases, among them the powerful Count of Monte Cristo. He takes revenge on everyone who conspired against him and is reunited with Mercédès.
5   Edmond reveals himself.

This list of events, or turning points in the A-plot, should already be part of the synopsis in the form of prose. This part of a synopsis must not be less than 20 pages long. Important secondary plots and their turning points are also part of a synopsis. Beyond this, a list of characters, their principal features and the main settings are also necessary, as are subplots, with characters and settings of their own. Subplots will be created, preferably with some connection to the A-plot. Additionally, depending on the telenovela's time slot, there must be at least one young character cluster and one comic nucleus. The A-plot usually has a sentimental and emotive tone, pleasing viewers who have been watching telenovelas for a long time and look out for the structure of melodrama; therefore, it is within subplots that other narrative genres often appear. In the case of *Totalmente Demais/Total Dreamer* (2016), the primary cluster was that of Eliza, played by Marina Ruy Barbosa; Jonatas, by Felipe Simas; Arthur, by Fábio Assunção; and Carolina, played by Juliana Paes. The amorous and family relations of these characters are also considered part of the A-plot. The secondary character clusters are often linked with the setting in which the main characters live or work. This strategy facilitates production by having specific actors for different sets, allowing more than one shooting crew to work simultaneously.

3.4 Strategies for developing a telenovela's step outlines and chapters and for relating to the production team

A telenovela's first chapter must introduce the protagonists and the detonating element of the A-plot's first plot twist, preferably through a critical event that will impart dynamism to the script. In Aguinaldo Silva's *Senhora do Destino/Her Own Destiny* (2004), for example, this event is the decision of the heroine to leave the *sertão* – the dry hinterlands of the northeastern Brazilian state of Pernambuco – with five children to look for her brother in Rio de Janeiro. Maria do Carmo and her children arrive in the city of Rio de Janeiro on 13 December 1968, on the exact day that the military dictatorship's Institutional Act no. 5 (AI-5) is being decreed – one of the most repressive acts of the military dictatorship. There is a significant disturbance in the city centre streets, filled with riots and police violence. Because of this critical event, the protagonist is separated from her brother and meets her antagonist, Nazaré, who will steal her youngest child. Aguinaldo Silva observes:

> When I created [*Her Own Destiny*], I decided to set the main character's arrival in Rio de Janeiro on that day, as she was experiencing all those terrible things that would mark her life. Without her realising it, a historical tragedy was taking place that would affect the entire country.
> (Fiuza and Ribeiro, 2008)

A telenovela's first week of chapters will establish the A-plot; viewers must understand and connect with the story.

Writing a telenovela is extenuating work given the volume of content involved. A telenovela's industrial rhythm leads each author to work out strategies to supply the demand for text production. Benedito Ruy Barbosa, for example, admits he cannot work with collaborators. Recently, however, he co-authored the telenovela *Velho Chico/Old River* (2016) with his daughter and grandson. Glória Perez works with more than one researcher, but also prefers to write alone and without a step outline. She picks up from wherever she left off the day before. In the case of working with collaborators, a step outline is essential. Through it the author will ask collaborators for scenes, dialogues or full chapters. Mauro Alencar analyzes how a telenovela step outline should be prepared: "It is all in the planning for the dramatic action over several chapters. In periodic meetings, the titular author and his collaborators define what will happen in the next chapters. Later, he details each sequence to be written and delegates tasks" (2002).

In *Totalmente Demais/Total Dreamer* (2016), after a weekly meeting with collaborators, quite similar to the one described by Alencar (2002), step outlines were prepared with scene descriptions and dialogue suggestions. The language is colloquial, and the instructions are for authors who are already very familiar with the narrative. Based on a step outline, Alencar (2002) describes three different ways of working with collaborators. In the first one, "by the chapter," different collaborators write chapters of the narrative by following the step outline. In a second way, "by character cluster or genre," collaborators write specifically for certain groups of characters or scenes with specific genres such as "action" or "humour." In a third option, "by scene or dialogue," an author may delegate only scene descriptions, parentheticals or dialogue.

Gilberto Braga works with a team of collaborators and has co-written more than one telenovela. In *Autores* (Fiuza and Ribeiro, 2008), he recounts how, in *Paraíso Tropical/Tropical Paradise* (2007), he used to hold weekly meetings with two of his collaborators to discuss the plot. He supervised the work of another writer (Ricardo Linhares), who was responsible for producing the step outline, or scene list, for each chapter that Gilberto Braga would later distribute to the different authors. Gilberto Braga explains the process this way:

> When I get the step outline, I review everything, any little thing I think I need to, and mark what each collaborator will be writing. I distribute the scenes to them. They write and then they send them to me. I put the chapter together and do the final edit.
>
> (Fiuza and Ribeiro, 2008)

Silvio de Abreu, who writes step outlines by himself, used to develop chapters with two collaborators according to the method Alencar (2002) describes as "by the chapter." Head writer Cao Hamburger, during his 2017–2018 season of *Malhação/Young Hearts*, discussed the step outline with the whole team and then followed the "by the chapter" method. On *Totalmente Demais/Total Dreamer*, the team of collaborators worked "according to character group or genre," but that did not prevent them from writing for other characters when necessary.

Since Glória Barreto and I co-wrote the telenovela *Malhação/Young Hearts*, season 2012, I have been perfecting a method of rolling out chapters that I use to this day. Every Saturday, the writer's team gets together to decide which of the chapter's cliffhangers would lead to the next turning point of the week and which subplots we will explore. These subplots may be responsible for a commercial break, but rarely for a cliffhanger. Commercial breaks are essential scenes that engender high expectations, but they are not as crucial as cliffhangers. Then the primary author or authors will prepare the step outlines of the week. The strategy is to craft the scenes in order to build the story leading up to each commercial break and, finally, to the cliffhanger. The break or cliffhanger scenes should not be isolated. Throughout the chapter, these scenes are constructed or, in the scriptwriter's jargon, "planted." David Howard and Edward Mabley (2002) call this scriptwriting technique *set-up and payoff* or *clue and reward*. Usually, near the resolution, when the characters' circumstances have changed, there is a "payoff" in which the gesture or action takes on a new meaning. The writer then chooses which scenes become clues in order for the plot to reach that point, usually the cliffhanger.

There is an average of ten scenes in a 43-minute-long chapter for a 7PM telenovela before each commercial break or cliffhanger, always alternating plots and settings. In a chapter with this running time, it is not unusual to offer, beyond the A-plot, two more subplots. The A-plot takes up at least half the scenes in the chapter that will total around 40 different scenes. The script may call for situations that are resolved within the chapter so that the viewer watching only that chapter does not feel lost in the narrative. In a telenovela, a script page is very similar to that of a film script; the difference is that telenovela's chapters usually contain more dialogue than a film, often underscoring the action.

The workflow plan I made for the script team of the three last telenovelas I wrote as the primary author runs like this:

SATURDAY

Staff meeting and elaboration of step outline 1, sent to the collaborators. The staff meeting discusses next week's main *breaks* and cliffhangers. These daily cliffhangers bear a direct relationship to the principal turning point laid out in the synopsis for that month and, consequently, with weekly subplot turning points.

MONDAY

The principal authors elaborate step outlines 2 and 3 and send them to collaborators.
(COLLABORATORS TURN IN SCENES FOR STEP OUTLINE 1)

TUESDAY

The principal authors elaborate step outlines 4 and 5 and send them to collaborators.
(COLLABORATORS TURN IN SCENES FOR STEP OUTLINES 2 and 3)

WEDNESDAY

The main authors elaborate step outline 6 and send them to collaborators. Principal authors reread and edit chapter 1.
(CO-WRITERS TURN IN SCENES FOR STEP OUTLINES 4 and 5)

THURSDAY

Principal authors reread and edit chapters 2, 3 and 4. (CO-WRITERS TURN IN SCENES FOR STEP OUTLINE 6)

FRIDAY

Principal authors reread and edit chapters 5, 6. Main writers revise and send the week's chapters (6) to production.

This continues for about six months, depending on the extension of the telenovela. Many writers prefer to begin with a large number of chapters already written. Silvio de Abreu used to work with 70 chapters in advance, for example, modifying them if viewers rejected a given character, plot point or actor. Unforeseen problems are commonplace: an actor might have an accident, get sick or the city may have a week of heavy rain, which could lead to script alterations. A well-wrought synopsis, however, with specific turning points, facilitates delivering chapters when the time allocated for the creative process is limited.

At Saturday meetings and in writing the week's step outlines, there is great care to restrict protagonists to six sets. Because there are six days of shooting per week, planning is needed to ensure that the script is viable for production. On average, at least 60% of a telenovela takes place in a studio, 30% on built exterior sets and 10% on location. In the studio, productivity is much higher, because of the number of cameras and lighting already designed and established by the set, compensating for the slower pace of a built city set and location shoots, which additionally suffer from every manner of unpredictable events: weather, traffic, construction, passersby.

A good relationship between the script crew and the director is essential, and practical examples abound. If an author does not communicate with the directors, an event may be impossible within the telenovela's time frame or budget, for example. By event, we mean a group of scenes that require investment. One example would be a chase scene with car crashes and stunt doubles; another example would be a party with a large number of extras; or an event that mobilizes most of the cast, preventing any possibility of shooting on other fronts – usually a wedding. The author must communicate in advance a turning point in the plot to the director because it may demand specific preparation for the actors. Also, the authors' use of songs, books and poems that need rights clearances must be anticipated. New sets with a significant number of scenes must be budgeted in advance lest they upset all production planning. In the telenovelas I worked in, we negotiated four-week intervals between each delivery of six chapters and the exhibition. This way we could still change the story if we needed to and the production had enough time for pre-production, one to two weeks, shooting, one week, and post-production, one week.

According to Alcides Nogueira: "When you write a page, you know it means work for three or four hundred technicians, production staff, directors, makeup artists, costume and set designers" (Fiuza and Ribeiro, 2008). Writing and producing a telenovela requires pacing, professional commitment and industry level resources.

The telenovela is a hybrid work with commercial characteristics and a business model based on advertisements. What I hoped to demonstrate by describing the elaboration of a telenovela narrative is that precisely because it is written while being exhibited, with an industrial pace of production, the primary author eventually retains autonomy over his creation. Oguri, Chauvel and Suarez (2009) interview several authors – Gilberto Braga, Manuel Carlos and Ricardo Linhares, among others – and TV Globo executives and conclude that, despite the search for knowledge through research, improvisation is an essential element of the production process. It focuses on the figure of the author, who transforms his or her original narrative according to his or her sensibility, whether influenced by what he or she has learned from research, which he or she may take into consideration or not, or by information gathered from his or her own everyday life. According to the researchers, "The telenovela writers interviewed in this study are almost unanimous in mentioning another source of information about viewers: remarks overheard on the street, from people known and unknown" (Oguri et al., 2009). To the researchers, to use a jazz band analogy, TV Globo executives lead but know that ultimately the musical result lies in the hands of their musicians: "And at Rede Globo, the great 'soloists' are the telenovela's director and, above all, the author" (ibid.).

Once a synopsis is approved, and the telenovela moves into the production stage, any manoeuvre demands the author's cooperation lest there be a break in the production mechanism and a significant loss of time and resources. In 2017, for example, according to data made available by the station during the 2017 International Emmy awards, TV Globo's telenovela chapter cost US$ 300 thousand per chapter at the time, including implementation, which means, in the case of *Totalmente Demais/Total Dreamer*, with 176 chapters, a total of US$ 52.8 million. The author, in the final instance, is the centre of a chain power of the most important and expensive Brazilian audiovisual product, that, beyond mobilizing hundreds of professionals and considerable resources, commands a massive viewership in Brazil. This power does not mean that telenovela authors do not worry about viewership or are unwilling to make changes to increase viewer numbers, a common interest in commercial television.

In the next chapter, I will reflect on the relationship between telenovelas, authors, broadcasters and viewers, taking into account the new technologies and transformations in spectatorship. I will discuss in what ways the connected viewer may or may not interfere in a telenovela.

Bibliography

Alencar, M. (2002) *A Hollywood brasileira- panorama da telenovela no Brasil*. Rio de Janeiro: Senac Rio. pp. 63–72.

Campos. (2007) *Roteiro de Cinema e televisão*. Rio de Janeiro: Zahar.

Comparato, D. (1996) *Da criação ao roteiro*. Rio de Janeiro: Rocco.

Dostoevsky, F. (1990) *The brothers Karamazov*. New York: Farrar, Straus and Giroux.

Dumas, A. (2003) *The count of Monte Cristo*. London: Penguin.

Field, S. (1995) *Manual do roteiro*. São Paulo: Objetiva.

Filho, D. (2001) *O circo eletrônico- fazendo TV no Brasil*. Rio de Janeiro: Zahar. p. 291.

Fiuza, S. and Ribeiro, A. (2008) *Autores: história da teledramaturgia*. São Paulo: Editora Globo. Vol. 1, pp. 44, 45, 48, 130, 224, 225, 414; Vol. 2, p. 214.

Howard, D. and Mabley, E. (2002) *The tools of screenwriting*. New York: St. Martin's Griffin.

Martín-Barbero, J. (1997) *Dos meios às mediações*. Rio de Janeiro: Ed. UFRJ. p. 174.

Oguri, L., Chauvel, M. and Suarez, M. (2009) O processo de Criação das telenovelas. *RAE*, Vol. 49 (1), pp. 44–46.

Perez, G. (2001) *O clone* (Synopsis).

Silva, C. (2010) *A corrida do ouro: o romantismo de Gilberto Braga*. M.A. Thesis, Pontifícia Universidade Católica do Rio de Janeiro, Departamento de Comunicação. p. 213.

Truby, J. (2008) *The anatomy of story: 22 Steps to becoming a master storyteller*. New York: Farrar, Straus and Giroux.

Vicentini, R. (2018) Walcyr Carrasco lembra quando se vingou de atriz que queria inventar falas. *UOL*. [online] Available at: https://tvefamosos.uol.com.br/noticias/redacao/2018/08/04/walcyr-carrasco-lembra-quando-se-vingou-de-atriz-que-queria-inventar-falas.htm?cmpid=copiaecola. Accessed: 12 December 2018.

Wilde, O. (2003) *The complete works of Oscar Wilde*. London: Collins.

4 The audience and the telenovela

Transformations and resilience of spectatorship

In Brazil, broadcast television attracts tens of millions of people daily to watch telenovelas between 6PM and 11PM, interspersed with local and national news. More than a device, television is a set of behaviours and practices, a pact, in continuous negotiation, with the audience. More than an offer of content, television is an arena and a starting point for dialogue with the viewer – and it is the viewer who holds the final word. In this chapter, the analysis of the telenovela will be from the public's perspective.

According to the research of the Ibero-American Observatory of Television Fiction by Vassallo de Lopes and Lemos (2019), in 2018 at least 60% of the audience was female in telenovelas aired from 6PM to 11PM on TV Globo. In that same year, 50.83% of the total population of Brazil was female. People from all socioeconomic backgrounds watch telenovelas, but a higher concentration of at least 48% of the audience is from C middle class.[1] All age groups watch telenovelas, but there is a higher percentage of people above 35. According to the research institute Kantar Ibope Media, in 2018, the 9PM telenovela had an average audience rating of about 28 million people per day in Brazil's 15 main markets. The percentage of women viewers is higher during 6PM telenovelas and higher among men at later hours. Regarding age group, the 7PM telenovela appears as the most viewed by children and adolescents. The 6PM and 9PM slots are the favourite of the public aged 50 and over. The programming of corporate television follows the daily life of Brazilians. Therefore, it is possible to affirm that the station produces content with the audiences described here in mind.

Eneida Nogueira (Svartman and Nogueira, 2018), research director of TV Globo until 2017, ponders that several reports point out television as a companion for the viewer generating a dimension of "belonging": the feeling that the viewer is part of society and is not alone since there is the notion of other viewers doing the same thing at the same time; television is a source of information to know what is happening outside the house; the next day, this viewer will have a repertoire in common with other people. As Wolton

(1996) observed, to watch a telenovela in Brazil is part of a social bond that mobilizes millions of people. However, the viewer adds to the narrative other secondary and tertiary texts, as Fiske (1987) theorized, in addition to personal experiences, expectations and dreams. The telenovela provides the Brazilian public with a reflection on its reality and current themes; provides information and knowledge; aids in identification with character traits, stories and emotions; offers an opportunity to fantasize and distance the audience from reality; offers relaxation and finally, inspiration. Almeida (2001) analyzed the consumption of the telenovela *O Rei do Gado/King of Cattle* (1997). She finds that, because it is about affective relationships, the telenovela interacts in these themes with spectators, provoking reflections on intimate and family relationships. Thus, the telenovela also constitutes a cultural text capable of promoting a singular sentimental education through a reflexive process of viewers with the narrative. Almeida (2001) cites as an example mothers who use the narrative to talk about delicate topics such as sexuality and love with their sons and daughters.

This chapter investigates the viewer's relationship with the telenovela under several aspects: viewer expectations and how the narrative adapts to them; the dialogue between the narrative and the viewer and where power lies in this relationship; and finally, how this balance is amplified (or not) by new technologies and interactive platforms.

4.1 What the viewer desires in a telenovela

Eneida Nogueira (Svartman and Nogueira, 2018) remarks that stations invest a significant amount of money on research to create linear programming. Mirian de Icaza Sánchez, also interviewed for this book, worked for 26 years at TV Globo Quality Central. She analyzed products according to their respective slots in programming until 2016. She worked alongside her late husband, Homero Icaza Sánchez, called "El Brujo" for anticipating trends in the 1970s and 1980s. She revealed that the department's research budget was, on average, US\$ 2 million annually at that time. "He – Homero Icaza Sánchez – even knew how long gossip took to go from Copacabana to downtown Rio de Janeiro" (Svartman and Sánchez, 2018).

According to Eneida Nogueira, 6PM is a time of transition. Different generations coexist there: the young have just returned from school, or have finished their homework and other activities, and that is the time that adolescents spend with their mothers and grandparents. "So, *Malhação/Young Hearts* – various seasons – was perfect because it fell well into their routine and the mothers', a time of conviviality with teenagers." *Malhação/Young Hearts*' story, cast and scenarios change every year, always approaching the young universe. This narratives' vocation fosters a great interchange between the women who are at home and their children, but at the same time, the content cannot embarrass either audience. Eneida (Svartman and Nogueira, 2018) exemplifies pondering a theme, such as sexuality. For mothers and grandmothers, sex is a health issue, and for adolescents, it is not. Besides, the young prefer to talk about sex with their peers and friends, not with family.

Mirian de Icaza Sánchez observed that the quality research department worked with vectors raised by the research to determine if a programme would be popular. "Homero said that when we received the audience rates, it was already too late. You need to know first what the trends are" (Svartman and Sánchez, 2018). It was Homero, with the certainty that they would be successful, who advised José Bonifácio de Oliveira Sobrinho, nicknamed Boni, then head of TV Globo, to produce period plots for the 6PM, nowadays 6:30PM, telenovelas. Nogueira (Svartman and Nogueira, 2018) comments that 6:30PM is when the woman is at home alone, either because the children and grandparents are busy with other activities or because the husband has not yet arrived home. It is a moment of intimacy in which this woman feels she can live inside herself. Narratives are fanciful and emotional, making room for period and romantic telenovelas. Eneida ponders: "Of course women change over time, and the telenovela has to change along with them. 6:30 p.m. is a time where she's making food, or preparing a snack, but it's a time of individualisation" (Svartman and Nogueira, 2018).

At 7PM, the house has a hectic dynamic: the husband arrives, the children and teenagers finish their activities, dinner is prepared and served. Therefore, the telenovela needs to follow this rhythm. "The telenovela sits at dinner table with the people," says Eneida Nogueira (Svartman and Nogueira, 2018). Moreover, because at 7PM a telenovela is seen by many children and teenagers, there can be no excessively violent or eroticized scenes.

Larissa Perfeito Barreto Redondo (2007) notes that, when addressing a mass audience, the 9PM telenovela seeks controversial themes and taboos that disturb emotionally and morally and that generate media about its themes. For Eneida Nogueira (Svartman and Nogueira, 2018), however, the 9PM telenovela has had the vocation to discuss society's great themes for the last 20 years. Nowadays, however, adults, children and adolescents are "in the room." Nogueira (ibid.) explains that one of the consequences is that the narrative can no longer be so dense. Through research, Nogueira concluded that this was a time for decompression for women who work outside the home and for homemakers. At 9PM they are exhausted. The researcher ponders, "This woman wants to reconnect with herself. It is the first moment she sits down and can afford to think something for herself. Learn, reflect, live something else" (Svartman and Nogueira, 2018).

Finally, the 11PM telenovelas are shorter, have more adult themes and more potent scenes. At this time, children have already gone to bed, adolescents prefer to be alone and adults watch these telenovelas in their bedrooms. According to Nogueira (Svartman and Nogueira, 2018), that is why authors can address topics such as violence and politics, and there are also more sex scenes. Because these are telenovelas that consider a more adult audience and, therefore, are less diverse, these works are the most similar to television series for a segmented audience. "11 p.m. is the time slot of intensity, and this can be in many ways. Women say it has to be something that the husband also likes to watch" (ibid.). Almeida (2002) notes that only a few telenovelas – usually those shown after the 8PM news slot – are considered relatively legitimate programmes for male audiences.

At first glance, Nogueira's analysis appears to take into account only the "traditional Brazilian family," which seems to be disintegrating today, along with other major political, social and cultural institutions. Muanis (2018) notes that historically networks organized programming according to an arbitrary, bourgeois and middle-class family model: conservative, white and heterosexual, ideally composed of a couple and two children inserted socially. He considers that this model gradually weakened in the United States as society transformed, women began to work outside the home, the pace of everyday life increased and, consequently, day-to-day schedules destabilized. According to Muanis (ibid.), another business model has emerged, with flexible schedules and diverse content for niche audiences – if we consider the new forms of hyper-segmented consumption and video-on-demand television.

However, what happened to broadcast television in the United States was not the same as in Brazil. Brazilian networks, as Nogueira (Svartman and Nogueira, 2018) observes, continues to adapt to the transformations of the Brazilian "model family." The objectives of broadcast television, when it elaborates audience research and fine-tunes its programming accordingly, is not to judge or impose a uninucleate family model, but, for economic interests, to please and conquer a massive audience. According to Martín-Barbero, "If television in Latin America still has the family as a basic unit of audience, it is because it represents for most people the primary situation of recognition" (1997). Even though there is ample content on different media platforms, the telenovela continues to attract millions of viewers, who engage, identify and are inspired by it; a different phenomenon from what occurred in the United States.

Audience surveys have as one of the vectors the number of connected televisions minute by minute. The *share* is the percentage of televisions tuned to a show. Research also points out the number of people who are at home, their gender, age, economic class and the percentage that watches the programme. According to Oguri, Chauvel and Suarez (2009), the National Television Panel (Painel Nacional de Televisão, PNT), used by research institute Kantar Ibope Media, employs the installation of *people meters*. This sample of the population is determined statistically according to sociodemographic parameters: "The main purpose of the PNT is to offer a description of TV consumption in the Brazilian market translated into information about tuning to different shows, home and individual viewing, penetration, profile, reach, and frequency of the exposure to the media" (Oguri et al., 2009).

At the end of the first month of a telenovela on TV Globo, the author and the director attend the first focus group an opportunity to fine-tune the dialogue with the audience. The researcher makes sure the group understands the plot, the characters, approves or not of love affairs or narrative tracks, understands or criticizes moral and aesthetic aspects of the story and aspects of production like costumes and sets.

The research company assembles groups that mirror the diverse audience at home during the time the telenovela airs. This group does not know what product they will review. The focus groups take place in São Paulo, not only because it is the largest advertising market in Brazil, but also because the average audience rates in the city are usually equivalent to or close to the average audience rates in the country. The researchers divide the audience between assiduous and sporadic viewers in addition to age, socioeconomic group and gender. The objective is to understand what pleases the faithful public and what is missing in the plot to attract the sporadic audience. When the audience of a telenovela is below expected, the station may decide to do more than one focus group, but usually one is enough to build a perspective of how the audience absorbs the narrative. After the qualitative research, it remains for the station executives, author and director to monitor the quantitative numbers daily and also the repercussions on social media – which do not mirror the total audience of the telenovela.

As an author, I have participated in four telenovela focus groups. Through a one-sided mirror, it is possible to see groups discussing the telenovela without being seen. To achieve the group's trust and complicity, the mediators talk about various subjects before entering the telenovela: daily life, children, work. In general, there are four groups per day for three days. In the groups of assiduous spectators, there is a greater desire to discuss minutiae of the plot, characters and romantic couples. In groups of sporadic viewers, it is more difficult to extract what they think, because they often only remember the plot in general lines and with various gaps. These groups know little about the narrative of the telenovela and prefer to watch something else, but they are the most important groups. They are potential viewers who are at home at the time of the telenovela, and we need to draw them in to watch the narrative.

The research experts will then elaborate, in detail, with the data collected, what pleases the viewers, what they dislike or even reject, including secondary plots. However, it is possible to get an idea of the results by watching the discussion groups – especially when there is rejection or affection for a plot or character.

Nogueira ponders: "It is rare for us to go to a focus group and find the author is surprised. They always have a slight idea of what is going on. I think it helps to organize the sensations, perceptions of the artists" (Svartman and Nogueira, 2018). Personally, I have never had a plot rejected, but some characters needed adjustments. For example, in one of the telenovelas I wrote, there were two best friends in love with the same character. However, the audience did not think he was worth it. We chose to bring qualities to the character, such as responsibility and solidarity, but we also created another male protagonist that entered the love triangle. In another telenovela, the adjustment was in a female character relevant to the plot, but with whom the women of the focus group did not identify; on the contrary, they found her harsh, very cold and arrogant. We then made her suffer all sorts of setbacks in a tide of bad luck and then transform because of this. The ratings rose significantly.

The author and the director are not obliged to modify the telenovela according to the research, but on commercial television low rates may mean that a telenovela will go off the air ahead of schedule or suffer from the intervention of an artistic supervisor; therefore, qualitative research can be a significant influence. It is also important to take into account that the Brazilian telenovela has commercial characteristics that permeate its production. Gilberto Braga ponders: "We do what the public wants if we agree with that" (Fiuza and Ribeiro, 2008). As an example, the author cites the research of *Vale Tudo/Anything Goes* (1988), in which the public asked that the character played by Regina Duarte, the heroine Raquel, not be so perfect. According to Gilberto Braga, the actress herself also wanted the character to have a slip. He attributes this desire to the Manichaean trait of the telenovela and the popularity of the villain, Maria de Fátima, daughter of Raquel, played by Glória Pires. Gilberto Braga says that he met with the other authors of the plot, and together they decided to keep the character with its original characteristics. "We let the public complain to the end because *Vale Tudo/ Anything Goes* was based on this opposition: honest mother and dishonest daughter" (Fiuza and Ribeiro, 2008). Gilberto Braga maintained the author's hierarchy as the base of the creative and power chain of a telenovela. The audience's acceptance or rejection is responsible for the transformations that the telenovela has been undergoing during the past seven decades. Nevertheless, the public expects a story well told by someone – the author – and not themselves. However, the relations between the telenovela, the author and the viewer have been changing along with society and access to new technologies, social tools and interactive platforms.

4.2 The repercussions of the telenovela and the influence of the viewer amplified by social media and interactive platforms

Even before the television set connected to the internet – or to the video game – the act of changing channels, turning switches on and off or performing tasks while a programme is being broadcast, among other practices, corroborates the arguments that the viewer is not and has never been passive. Moreover, there is no passivity in a viewer that relates the television text to their own experiences, readings and within their historical and social context, because every interpretation process is subjective and active. The understanding of an active audience can be mistakenly associated only with the fan figure, with the notion that this is the viewer who interacts with the work ostensibly, producing content, for example. As already observed, the active audience engages with the story emotionally, associating the telenovela, for example, with secondary and tertiary texts and interacting in various dimensions. Fiske (1987) remarked that there is a negotiation process between the content proposal and the viewer's position. For him, in this negotiation, the power lies with the public.

TV Globo's Customer Service Center (CAT) was created in the 1980s. Until 2020, the viewer could contact TV Globo directly by telephone; only after 2000, by email as well. According to Oguri et al. (2009), among the viewer's suggestions are errors in scripts or production and the inadequacy in the approach of topics connected to associations or entities. CAT also receives public statements with criticisms, compliments, suggestions, complaints and questions and for varied information about the programming schedule. The vast majority of viewers interact through social media. Since 2011, TV Globo has kept a corporate page on Facebook, www.facebook.com/RedeGlobo/, with the following message:

> This page is a place for our audience. Comments, suggestions, criticisms, and compliments are welcome. We need, however, to have certain rules. We will not accept spam, chain letters, or inappropriate content. We also reserve the right to remove any posting or other inappropriate material.

The page had approximately 14 million followers as of June 2020. The network has had a profile on Twitter since 2008 with 12.3 million followers,

2.81 million subscribers on YouTube and 10.4 million followers on Instagram, all as of June 2020.

Nevertheless, until 2020, the station maintained the call centre for those who still preferred this means of contact. These interactions do not appear publicly on social media profiles of the broadcaster; but CAT needed to share the information. During the 2015–2016 *Malhação – Sonhos/Young Hearts – Dreams* season, CAT contacted us, the authors, to report a message that had thrilled the attendant and her supervisor. The message was about a plot, secondary to the main story, of a severe and determined teacher named Lucrecia, who raised her troubled teenage daughter alone and discovers she has breast cancer, and an unexpected fragility takes over her life in an overwhelming way. For a few months, the narrative followed the discovery, denial, treatment and overcoming of the character and her family. The fight against the disease throughout the season had numerous reactions among spectators. Here is the message from a viewer received by CAT on 10 April 2015:

> I am a big fan of Young Hearts! Yes, I am 26 years old, and I've seen every season, even with criticism, for some say I'm too old to watch. Well, I have seen all the stories possible, but today's scene was profound for me, even though it was already a subject spoken of earlier. It's been 17 years since my mother had cancer, and a year ago it came back hard. Four months ago, my mother had surgery to remove her left breast; it was a heavy day. And today, I revived, in Young Hearts, that day but in a lighter way. Anyway, the authors, as much as they have already heard this, should be happy to be able to portray something so painful in a light way and with endless love. Congratulations to the artists who gave their souls to this role, because it is necessary. They say that "art imitates life," but well, I think it is art that helps us face life.

This message is an example of what Fiske (1987) analyzed as a vertical intertextual relationship, in which, starting from the primary text of the telenovela, a viewer produced a tertiary text relating the narrative to her personal experience. As an author, I recorded a statement to be sent to the spectator, as did Helena Fernandes, the actress who played Lucrecia, thanking her for the message. Besides, we paid a quick tribute in one of the characters' scenes, quoting part of the message as if it were from a friend of Lucrecia's.

However, another scene generated more reactions in the plot of Lucrecia's struggle against breast cancer: the one in which actress Helena Fernandes does a breast self-examination, naked from the waist up, in front of the mirror. We planned the scene to air precisely in October, when the worldwide movement Pink October occurs, to raise awareness of the importance of early detection of breast cancer and to share information about the disease. The comments on social media during the programme reflected extreme and opposing positions. While some criticized or mocked nudity in the late afternoon in a telenovela for a young audience, others supported the initiative. Also, perhaps due to the reactions on social media, this scene got a high response on blogs and websites specializing in television criticism, texts that would be, according to Fiske (1987), secondary. Among them, Notícias da TV/TV News, a website linked to the UOL News portal, one of the main and most popular in the country, published an article entitled "Globo displays breasts in *Malhação* at 17h24 and shocks viewers" (Castro, 2014). In the article, Daniel Castro writes that the scene "impressed viewers on social media," but acknowledges that the indicative rating allows nudity in the afternoon provided it has no sex appeal. The article Daniel Castro wrote on the scene of the self-exam in *Malhação*/*Young Hearts* is open to comments: the primary text, the telenovela, led to the secondary text, criticism, which provided an encouraging environment for tertiary texts, which are the viewers' comments. To participate, one must register on the website, share personal data and supposedly publish their real name. Nevertheless, many participants use fictitious names, circumventing the identification system. Furthermore, as the article was only published the following day, the comments were not written directly after viewing the scene, like on Twitter. Even though the answers are more elaborate, there was also a polarized response, as on Twitter: Valmir Fabio Versolato, who identifies himself as a lawyer and musician, found it absurd that nudity still shocks anyone in the 21st century; Daniel Lindenberg, from the northern state of Piauí, argued that teenagers, in his opinion the primary audience of the programme, do not care about breast cancer and that they probably had another reading of the scene of the self-exam; Tina Oliveira Bortuluci revealed that she had breast cancer, and discovered it precisely by doing a self-exam. She made a point of clarifying that, for her, it was not a sex scene. However, Daniel Lindenberg pointed out that if a woman showing her breast is allowed to be a campaign to prevent cancer, then "How about a man examining his PENIS for prostate cancer at five in the afternoon in *Young Hearts*? It is for a good cause."

Studying this vertical intertextual production is a way to access meanings of this circulation and mobilization. For example, why does this scene – which supposedly contains no sexual bias – bother people so much? Because of the interaction between the telenovela and society and the conflicts of the present day. Morals, ethics and political polarization are some of the issues that arise. As Sibilia (2008) ponders, what is obscene in nudity also changes according to the historical moment, and today we witness conflicting forces and movements of advancement and setback. Kenneth Clark (1987) proposes that there are necessarily two different ways of understanding and defining nudity. *Nude* would be the nudity of the ideal form, the classic inspiration of the Greek model. *Naked*, for Clark, would be the standard, bare, artistically unrepresented human body. It is the obscene condition of nudity.

Even though the self-exam scene of this narrative had more reactions on social media and specialized blogs, as an author, I only chose to absorb the viewers' contribution through CAT into the telenovela. Up to this day, in a telenovela on TV Globo, unless there is a real error, for example, a wrong credit for a song, or risk of prosecution with solid reasons, or to the indicative rating, the author has the final choice.

Fiske was Henry Jenkins' professor. Jenkins is one of the leading authors discussing the growing power of participatory television audiences, whose opinions and productions circulate and resonate in social media (Jenkins, 2006; Jenkins et al., 2013, 2016). According to Jenkins, Ford and Green (2013), the new interactive tools and platforms enable audiences to consume content initially produced for television and produce new content from it. The viewer currently manifests his or her opinion about audiovisual content in different social media, participates in discussion groups, and produces content derived from characters and narratives with unprecedented immediacy. The question is whether these opinions rhave had a higher resonance recently to the detriment of the influence the viewer has always had on telenovelas, expressed through focus groups or ratings, or access to customer services.

For Jenkins, Ito and Boyd, the content-producing fan universe is born out of fascination and some frustration: "If you weren't fascinated, you wouldn't continue to engage as a fan. If you weren't frustrated, you wouldn't continue to rewrite and reinvent" (2016). For Lopes and Lemos (2019), it is from the moment that the viewer begins to get emotionally involved with the plot and to create deep bonds with fiction that he becomes a true fan: "This fan will tend to explore as much as the production offers, he will know the characters and the direction of their stories" (Lopes and Lemos, 2019). For these authors, at some point, the fan becomes a producer himself when he realizes that the plot can be expanded, either through his personal experiences or shared experiences in fan communities and social media. The production of content by fans, the amplification of the old "word of mouth," with circulation in social media and interactive platforms is a phenomenon of the present, but academics differ on the real power of this connected audience.

Exchanges between the viewer and the broadcaster – or the content producers – have always existed through the history of television, but social tools have deepened and amplified this practice. By researching public participation in Brazilian telenovelas since the 1950s, using specialized magazines and letters to TV stations and actors, Baccega et al. observe the transformation of the telenovela's audience:

> In a way, the study commented here confirmed the assumption that blogs, social media, and other digital environments would represent a space of expansion of behaviour that began at a time when readers columns in specialised (print) magazines and fan-clubs were already crucial mediators of the fictional-symbolic/daily/imaginary relationship.
>
> (Baccega et al., 2015)

Baccega et al. (2015) also study the supposed increase in the importance of the viewer in Brazilian television. For researchers, today's audience acts as a consumer of media and products marketed along with narrative content through advertising or merchandising. The relationship between fan cultures and consumer culture mechanisms is also observed by Hills (2002), as fans are always consumers. The scholar notes that fans are no longer seen as annoying eccentrics, but rather as loyal consumers to be courted. The definition of a fan differs from author to author, and Hills (2002) refers specifically to fans of serials with fragmented audiences. In this book, we consider fans to be all those who maintain an affectionate relationship with the work; it is necessary to note that only a portion of these fans devote time to circulating their opinions on social media and producing content for participatory platforms.

Shirky (2010) builds a paradigm through his research of this fan participation. It is a pyramid in which few produce much content for many, contrasting with the concept of a more horizontal participation culture proposed by theorists such as Jenkins (2006), Jenkins et al. (2013) and Jenkins, Ito and Boyd (2016). For Jenkins et al. (2016), the culture of participation is one in which democratic values and diversity in all aspects related to the interaction between participants are accepted, taking into account that we are all able to make decisions, collectively or individually, and that we can express ourselves in different forms and practices. It is a culture with few barriers to expression, artistic production and engagement, which supports the creation and sharing of this creation, and a balance in which the most experienced teach the less experienced. Members of participatory culture believe, according to Jenkins, that all contributions are significant and that there is a social connection between them. Regardless of the degree of participation, given that advertisements support the business model of corporate television, the audience has always been relevant; however, the significant dedication of some fans who amplify the potential for the resonance of a show naturally interests broadcasters. Jenkins et al. (2016) note that there is a tension between this fan culture and the industry, from which fans get the content that is relevant to them. For the authors, fan communities and other content producers are struggling to gain more access to distribution and circulation.

There are several fan blogs and profiles on different social media dedicated to the narratives of telenovelas, recording new stories, characters and events and comparing them with the previous ones. Baccega et al. (2015) note that most of them remain amateurs, although they may have tens of thousands of followers. Telenovelas have always received attention from the specialized media in Brazil, but social tools increase the importance of amateur commentators. Sérgio Santos, for example, is a biologist who is also an amateur critic, and his Twitter profile (@zamenza) has approximately 84,000 followers as of June 2020. He also writes for his blog and keeps profiles on other social media without as much popularity. Recently, fan clubs and amateur critics like Sérgio Santos are part of the communication strategy for the launch of a new telenovela. They can be invited to a press conference or be part of an online event.

Fans produce content because of the affection they have for the work, and it is this affection that makes them seek more content produced by their peers and also by the station. Most content produced by the station requires little public participation, but there are several examples of transmedia actions in which public participation is essential: either in voting or in the production of content for the programme and crowdsourcing.

About this creative ecosystem around the telenovela, Fechine and Figueirôa observe: "The interactional universe triggered by the project is not limited to the strategies proposed by the producers and, therefore, is not entirely under their control" (2015). This universe would involve both actions that are either an expected "response" of consumers to the convocations of producers and activities that, coming from these consumer-producers, are unexpected and even "deviate" from their objectives. The telenovela *Deus Salve o Rei/God Save The King* (2018) by Daniel Adjafre, for example, invited fans to participate in the production in various ways: by visiting the studios, getting to know the actors, being part of workshops about the show and even entering the scene as extras (Gshow, 2018). There was a tacit understanding between the network and these fans that they would produce content from the experience that would help promote the telenovela. There is no way to control what fans share; however, they knew that the broadcaster could deny them that opportunity in the future. According to Jenkins, Ito and Boyd, producers seek to control this engagement for their interests. "Participation implies some notion of affiliation, collective identity, membership, but beyond that, we have much to figure out if we're going to continue to apply this framework to contemporary digital culture" (Jenkins et al., 2016). The use of a large broadcast of content produced by fans is questionable ethically, since, on the one hand, we have amateurs and on the other hand, corporate companies. However, both get what they want: the network gets publicity and fans use content that belongs to the station, such as characters and plots, as raw material for their independent productions.

Although the audience is a guide to the longevity and integrity of a telenovela, new social tools and technologies that increase the circulation of content produced by producer-fans – a minority – do not currently influence the massive audience of daily television dramaturgy. Thus, in the specific case of the telenovela, they do not exert a significant influence on the product. However, this does not mean that these contributions cannot enrich the telenovela or that this scenario cannot change if the daily drama becomes a product of a fragmented audience and the television business model changes.

In their chapter in a book coordinated by Maria Immacolata de Vassalo Lopes, Baccega et al. conclude that "[w]hat becomes increasingly clear is the importance and, we would say, the strength of the viewer-receiver, even if eventually only a fan in the economy of the cultural media industry" (2015). On the other hand, Stycer (2016), *Folha de S. Paulo* researcher and critic, considers that this research by Baccega et al. is limited, pointing out that "there is still no way, in the digital universe, to have a clear idea about who are the participatory fans, those who vote in polls, post comments on blogs, share links on social media." He notes: "there are different groups, with multiple interests, acting amid digital anonymity. In the case of television programming, it is interesting to notice the movements that these fans (or 'militants') make both in defence of each other and the effort to disqualify rivals" (Stycer, 2016). Thus, he justifies the fact that young actors, popular on social media, get votes in awards with a popular vote. It is the mobilization of fan clubs and not a picture of what the "people" think. "The results, different from each other, have something common: they should not be deemed sanctified as 'the voice of the people,' but rather as the voice of 'some people' or 'some peoples'" (ibid.).

The difference between these actions, polls and formal audience surveys is that in the focus groups, ratings are based on the whole audience that watches the telenovela, a thermometer of commercial sales and the advertising-based business model. This audience is not the same as the one that votes in social media polls or that produces content. For Eneida

Nogueira, the reactions on social media does not replace qualitative and quantitative research:

> Whoever is talking in the social network are the ones that either love or hate [the programme]. However, the group that does not manifest itself so clearly is the group that participates in the research. I think it is important to listen to these people because they are the vast majority.
>
> (Svartman and Nogueira, 2018)

In Brazil, the reactions on social media do not mean large audience numbers on open television. Programmes such as reality shows *The Farm/A Fazenda* and *Masterchef Brasil*, for example, attracted much attention on social media in 2019, including several trending topics. However, this success on the internet did not necessarily revert to audience ratings on open television. This difference makes sense if we consider that the penetration of open television is higher than that of the internet in Brazil.

According to the annual study of the GMSP (Media São Paulo Group, www.gm.org.br), in 2018 in Brazil, there were 68,920,836 households with a television set, amounting to 96.8% of the population. Geographically, TV Globo's signal covers 97.2% of all households with a television in Brazil. Between 6PM and 10PM, when telenovelas are being aired, approximately 70% of these televisions are on. Also, according to media measurement and analytics company Comscore, in 2018, 46,170,641 Brazilians accessed social media networks. Facebook had 55.7% penetration within these users, while Instagram and Twitter only 16.3% and 10.4% each. Therefore, although these numbers are possibly changing, the visualization of massive audiovisual content still occurs by open television in Brazil. Another point is that the television metrics that guide the price of advertising consider only the 15 major cities to reach the PNT. Many of the people that may vote in an internet poll or be part of an internet campaign do not live in these cities. Once again, it is necessary to take into account the business model of open television, supported by advertising since the beginning, which makes ratings a vital factor in curating the flow of programming.

For digital platforms, for example, a programme with enormous resonance on the internet can be attractive. That is because a fragmented, specific and segmented audience consumes this content. In the case of digital platforms with subscription-based business models, it is the catalogue that needs to be attractive to the consumer in the form of a database. The reaction on the internet becomes equivalent to promotional activity. The risk of rejection is diluted not only by the number of titles in a vast menu of content but also by the platform's algorithm, which supposedly offers what the viewer wants. Therefore, as the algorithm fits this consumer's taste, fewer rejected programmes will be suggested to him. It is not the massive audience of the programming flow of corporate television that supports the business model.

Another critical factor is that broadcasters have already learned that fans of certain characters or actors know how to circumvent the algorithm of some social media platforms to draw the attention of the production or its authors. On Twitter, for example, many users have already realized that if they commit to posting the same hashtag at the same time, it will show in the trending topics. Twitter's algorithm "understands" that it is an important subject, and the "headline" will appear in the featured list of trends of the moment. Currently, it is up to the author to follow (or not) the suggestions of research and social media, even knowing that these do not mirror the actual audience of a telenovela.

As journalist and critic Maurício Stycer remarks, "[c]urrently celebrities, including telenovela actors, have also become powerful media on social media. To get more followers and thus more advertising proposals, they need to interact with the audience and become characters of their own private life" (2019). Thus, a fan-coordinated action can directly target these actors while they are on the air in a telenovela. This does not necessarily mean that it affects the audience of the telenovela, but these actors may try to pressure the station for changes. Also, the mismanagement of a crisis can cause repercussions to blow out of proportion. However, these crises have already brought practical experience to corporations. In the article, Stycer (ibid.) analyzes the conflict between the actors of telenovela *Sétimo Guardião/Seventh Guardian* (2018/2019) and their fans that was all over the media and, of course, their social networks. Author Aguinaldo Silva decided to respond with a public statement ensuring that the narrative would not change because of that. There is also the practical learning progress of the actors and entrepreneurs who represent them since the business model involving social media is part of the income of these artists.

In the next chapter, the business model of corporate television, which directly influences the telenovela, will be addressed. The chapter analyzes the new possibilities and challenges that arise with digital platforms and the sliding of content between different screens and media. If the telenovela in Brazil has as its characteristics longevity and massive audiences, it only continues to exist because it provides profits to its networks.

Note

1 According to the Brazilian Association of Research Companies (ABEP), the Brazil Economic Classification Criterion emphasizes its function of estimating the purchasing power of urban people and families, abandoning the claim to classify the population in terms of "social classes." The division of the market by economic

classes: www.google.com/url?sa=t&rct=j&q=&esrc=s&source=web&cd=10&
cad=rja&uact=8&ved=2ahUKEwiJ9sCGnoXgAhU7JrkGHfVJDYoQFjAJegQ
IBhAB&url=https%3A%2F%2Facademia.qedu.org.br%2Fglossario%2Fnivel-
socioeconomico-nse%2F&usg=AOvVaw1l5S3l9hhUe5zv4ba3hjBr. Accessed:
19 January 2019.

Bibliography

Almeida, H. (2001) *"Muito mais Coisas": Telenovela, Consumo e Gênero*. Ph.D.
Thesis, Universidade Estadual de Campinas, Instituto de Filosofia e Ciências
Humanas.

Almeida, H. (2002) Melodrama Comercial: Reflexões sobre a Feminilização da Telenovela. *Cadernos Pagu* (19).

Baccega, M., Tondato, M., Orofino, M., Nunes, M., Junqueira, A., Budag, F., Abrão,
M. and Marcelino, R. (2015) Fãs de telenovelas: construindo memórias – das
mídias tradicionais às digitais. In: Lopes, M. (ed.) *Por uma teoria de fãs da ficção
televisiva brasileira*. Porto Alegre: Editora Sulina. pp. 99, 102.

Castro, D. (2014) Globo exibe seios em Malhação às 17h24 e choca telespectadores.
Notícias da TV. [online] Available at: http://noticiasdatv.uol.com.br/noticia/televisao/
globo-exibe-seios-em-malhacao-as-17h24-e-choca-telespectadores-5122.
Accessed: 8 August 2016.

Clark, K. (1987) El Desnudo corporal y el desnudo artístico. In: *El Desnudo: un
estudo de la forma ideal*. Madrid: Alianza. pp. 17–39.

Fechine, Y. and Figueirôa, A. (2015) Transmidiação: explorações conceituais a
partir da telenovela brasileira. In: Lopes, M. (ed.) *Ficção televisiva transmidiática no Brasil: plataformas, convergência, comunidades virtuais*. Porto
Alegre: Sulina. p. 325.

Fiske, J. (1987) *Television culture*. London; New York: Methuen.

Fiuza, S. and Ribeiro, A. (2008) *Autores: história da teledramaturgia*. São Paulo:
Editora Globo. Vol. 1, p. 416.

Gshow. (2018) *Fãs assistem cena de Afonso e encontram Romulo Estrela em
workshop de 'Deus Salve o Rei'*. [online] Available at: https://gshow.globo.
com/novelas/deus-salve-o-rei/noticia/fas-assistem-cena-de-afonso-e-encontram-
romulo-estrela-em-workshop-de-deus-salve-o-rei.ghtml. Accessed: 19 January
2019.

Hills, M. (2002) *Fan cultures*. London: Routledge.

Jenkins, H. (2006) *Convergence culture: Where old and new media collide*. New
York: New York University Press.

Jenkins, H., Ford, S. and Green, J. (2013) *Spreadable media: Creating value and
meaning in a networked culture*. New York: New York University Press.

Jenkins, H., Ito, M. and Boyd, D. (2016) *Participatory culture in a networked era*.
Cambridge: Polity Press. pp. 66, 67.

Lopes, M. and Lemos, L. (2019) Television distribution models and the internet:
Actors, technologies, strategies. In: Lopes, M. and Gómez, G. (eds.) *Ibero-
American observatory of television fiction*. Porto Alegre: Sulina. p. 19.

Martín-Barbero, J. (1997) *Dos meios às mediações*. Rio de Janeiro: Ed.
UFRJ. p. 293.

Muanis, F. (2018) *A imagem televisiva- autorreferência, temporalidade, imersão.* Curitiba: Appris Editora.

Oguri, L., Chauvel, M. and Suarez, M. (2009) O processo de Criação das telenovelas. *RAE,* Vol. 40 (1), p. 43.

Redondo, L. (2007) A telenovela brasileira: uma apresentação de seu formato e de seus aspectos principais. *Cenários da Comunicação,* Vol. 6 (2).

Shirky, C. (2010) *Cognitive surplus: How technology makes consumers into collaborators.* [ebook] New York: Penguin. Accessed: 10 June 2010.

Sibilia, P. (2008) O corpo reinventado pela imagem. *Trópico.* [online] Available at: http://p.php.uol.com.br/tropico/html/textos/3030,1.shl. Accessed: 4 November 2008.

Stycer, M. (2016) *Adeus, controle remoto: uma crônica do fim da TV como a conhecemos.* Porto Alegre: Arquipélago editorial. p. 26.

Stycer, M. (2019) *How Instagram went from artificial paradise to celebrity hell.* [online] Available at: https://tvefamosos.uol.com.br/blog/mauriciostycer/2019/02/22/como-o-instagram-foi-de-paraiso-artificial-a-inferno-das-celebridades/?cmpid=copiaecola. Accessed: 20 February 2019.

Svartman, R. and Nogueira, E. (2018) Interview with Eneida Nogueira.

Svartman, R. and Sánchez, M. (2018) Interview with Mirian de Icaza Sánchez.

Wolton, D. (1996) *Elogio do grande público – uma teoria crítica da televisão.* São Paulo: Ática Editora.

5 The influence of screen convergence, new digital platforms and transmedia narratives on the telenovela business model

The Brazilian telenovela is a commercial product of broadcast television that has a well-known business model: the audience combined with the credibility of the audiovisual product that establishes a value for the commercial exploitation of a given work. Not every programme with an expressive audience attracts advertisers; this is the case, for example, of news or shows with excessive violence or sexual content. As a production premise, telenovelas have to be attractive to the audience and also to advertisers who finance the content, generating profits for the company. Since the beginning of television networks in Brazil, daily television dramaturgy has had this vocation. If in the beginning advertising agencies supervised telenovelas and were responsible for spaces in the television grid, now the business model works the other way around.

Robert Allen (1992) notes that being a spectator of commercial television entails an implicit contract. The narratives resemble a "gift" to the audience, as the open TV signal arrives unsolicited and free in viewers' homes. Television commercials are something that come with the "gift," but are also a reminder that an advertiser is the bearer of such a "gift." Allen perceives that at least some viewers are expected to purchase the advertised product; therefore, the conditions of the implied contract will not be carried out in front of the television set, but in a store or supermarket. Allen compares the implied contract of television with that of cinema. For him, the audience of a film would not accept commercials because he or she has paid for the ticket, and no other action would be expected of them. Nevertheless, nowadays, with the convergence of the screen business model mix, the separation between media is not the same as it was in 1992, the year the article was published. For example, it is currently common to watch advertisements before the main feature in cinemas, and films themselves are a means of marketing products.

According to market research company Nielsen, while *Star Wars: The Force Awakens* (2015) grossed US$2 billion in theatres, *merchandising* revenues were US$5 billion to US$6 billion (Nielsen Research Institute, 2015). In 2016, Mattel, the brand of the year according to the MIPTV audiovisual industry event, presented in Cannes its content division responsible for producing films and creating narratives from their toys (YouTube, 2016). The company shared the information that, with the drop in sales, creating narratives for various media and screens, including films, has become critical to selling toys. Throughout this chapter, we will analyze the different business models of audiovisual content and how they currently converge in the media ecosystem.

By including the financial aspect in his theories about new media and interactive platforms, Scolari (2009) defines a *media ecosystem* as the relationships between the media and the economy and the perceptual and cognitive transformations suffered by individuals exposed to communication technologies. It is a broader definition of convergence in contemporary times, which also encompasses serial narratives and their relationships with the public. This new phenomenon described by Scolari (2009) brings challenges to the future of audiovisual content, since each screen or platform has its characteristics of spectatorship, consumption and interaction and a different model of financial return.

Audiovisual producers and distributors plan content for various screens, possibly in different formats, with the purpose of higher financial gain. The strategy of sliding between screens not only increases the economic return of the works but also meets the public's increasing cross-platform demand. A few years ago, a film would have the cinema as its first window, then DVD, VoD – video on demand – later cable television, and then open television. However, Alfonso Cuarón's *Rome* (2018) was produced by Netflix for the digital VoD platform and screened in 600 theatres around the world, a limited distribution. Most of the screenings were free and promotional, and Netflix's interest was primarily to leverage the film, included by the American Film Academy in the list of films that competed for the Oscars in 2019. *Rome* was a finalist in ten categories, winning the award for best director, best foreign film and best cinematography. Interestingly, in an interview with *Variety* magazine, the director, Alfonso Cuarón, said that the best way to watch his film was in cinemas (Variety, 2018). The pandemic in 2020 accelerated this tendency of not necessarily respecting the order of screens for film content, as films made for the cinema, like *Trolls 2*, went directly to VoD. NBCUniversal CEO Jeff Shell (D´Alessandro, 2020) admits that there is a growing segment of the population that does not go to movie theatres. The telenovela's distribution and business model is also changing due to the contemporary media ecosystem. During the pandemic, digital platform Globoplay announced 100 new telenovela titles from the TV Globo collection in its catalogue, giving this content new distribution and commercial possibilities.

When analyzing tendencies and business models, it is also necessary to ponder on transmedia content and narratives. For Jenkins (2006) and Jenkins, Ford and Green (2013), these would be those that represent a process in which elements of a fictional universe are systematically dispersed in various channels for the display, circulation and distribution of content to create a unified entertainment experience. Ideally, each media contributes uniquely to the narrative. Jenkins (2006), however, notes that there is a kind of "mothership" of the narrative, the first work of a franchise that will be developed, for example. Exhibition of the same original content in various media is not considered a transmedia narrative experience. Scolari (2015) defines transmedia narrative as a particular narrative structure that expands through different languages – verbal, iconic, for example – and media – cinema, comics, television and video games. The different media and languages participate and contribute to the construction of the world of transmedia narration.

For Robert Pratten (2012), in a practical guide for the production of transmedia narratives, this type of narrative makes it possible to tell the same story through multiple media and, preferably, with a degree of public participation or collaboration. Engagement, according to the author, increases audience satisfaction and affection for the universe/story. For him, the story needs to be greater than the sum of the narratives in different media, so that the viewer's experience will be richer and more exciting when he accesses the total creative universe of the narrative. The author also offers a business model design in which different transmedia narratives slide through various media, sometimes as a promotional action, sometimes with the possibility of financial return.

Conversely, Mittell (2015), writing about the narrative complexity of American TV serials, notes that these narrative extensions are not a new phenomenon. Even if the term is relatively new, the strategy of adapting, expanding and enriching the narrative in other media is as old as media itself. The author cites as examples paintings based on passages of the Bible and fictional characters of the 19th century such as Frankenstein, whose trajectory has already been narrated, transformed and extended in different narratives in various media. For Jenkins et al. (2013), however, there is a big difference between extending a narrative *to* other media and reproducing a narrative *in* other media. An adaptation is a way of retelling the story in another media; an extension seeks to add something to the story that already exists, as it slides from one media to another.

From a first text or "mothership," several secondary transmedia narratives can be offered by the producers of a television programme on the product-related website. As Pratten (2012) has observed, these transmedia narratives add value to the product concerning audience affection, demand

and satisfaction. An example of this transmedia narrative in a telenovela may be a vlog on a specific character or a web series, in which secondary characters may expand a plot of the story. An example is Orelha TV, a transmedia experience in the 2012–2013 *Malhação/Young Hearts* season, in which I was head writer, and which was nominated for the 2014 Digital Emmy. Orelha TV – Ear TV – is an internet video channel of one of the characters in the plot. The images of Orelha TV were captured during a scene by the actor who played the character, David Lucas. In the programme, small excerpts of this material appeared, but viewers could follow the Orelha TV channel in full on the internet.

In the case of the telenovela, the first text which airs will be more important to the broadcaster than the transmedia narratives, even if they extend the narrative and contribute to the fictional universe. It would be a transmedia narrative structure with an explicit centre and a set of *satellite narratives* that support it. Jason Mittell (2015) notes that in the case of television, commercially, it may not be attractive for specific programmes to invest financially in multiple platforms because most of the audience may not be interested in more than one media.

Shirky (2010), however, considers that it is not only the desire for more information that makes a fan or viewer seek transmedia narratives, but the affection for the original work. The viewer wants more. In the case of the telenovela, the vast majority of viewers will be involved only with the main narrative. The telenovela has a pleonastic language that takes into account the fact that the viewer may have missed one or more chapters. It is the affection for the story, for the characters, that can generate demand for more content, without interfering in the enjoyment and experience of watching the telenovela on the TV set within the linear flow of programming.

However, there is another demand for transmedia narratives that is not one of more information or affection. As Pratten (2012) has pointed out, there may be a transmedia narrative that brings a financial return. A transmedia narrative can complement the business model of a television programme. The telenovela I co-wrote, *Totalmente Demais/Total Dreamer* (2016), for example, had a spin-off series sponsored by a brand of beauty products, which will be a case study in the next chapter.

It is necessary to understand media convergence from the perspective of content and of the business model. This chapter discusses the relationship between the telenovela narrative, its consumption, the audience and the telenovela's primary business model: advertising. This chapter also discusses how this relationship can be transformed with content sliding between platforms and transformations in spectatorship.

5.1 Free content

Up to this day, the telenovela has a traditional business model, supported by the audience and credibility of the work to sell commercial spaces. Simultaneously watching an original chapter on television with most of the country, time zones considered, is part of the value that the telenovela offers. However, publicity suffered an 8% drop in 2019 in relation to 2018, according to the Globo Group 17 May 2020 report in the newspaper *Valor Econômico* (Valor Econômico, 2020). As this book is being published, we are in the midst of the COVID pandemic, where the resulting drop in publicity may be even more significant. On the other hand, Globoplay, the digital content platform of Globo Group, had a 55% increase in profits from 2018 to 2019 (Jardim, 2020). As of September 2020, the platform offers and streams content from cable stations of the Globo Group as well. Information released by Globoplay before and during the COVID pandemic states that telenovelas, as a whole, never left the top ten most-seen contents of the platform, which includes original and international series and films. Therefore, it was the platform's decision to provide 100 telenovelas from the TV Globo collection as part of their catalogue during the pandemic in 2020. Globoplay already provides online streaming of TV Globo linear programming for free, but once telenovelas prove to have value as part of the menu offered for subscription, a new way to profit with this content will emerge and free sharing of chapters by fans may become a problem. Also, according to Kogut (2020), Viva, a cable television channel that reruns telenovelas, leads the audience in cable TV. "Nostalgic people celebrate. Fan clubs are also resurrected (in social media). Everyone knows the outcome of these stories, but that does not diminish their power to attract viewers."

Most television stations use content identification tools, such as *video fingerprinting*, to remove pirated material from free sharing sites. The goal is to concentrate and control the audience in display windows planned by producers, exhibitors, distributors and corporate television. It is a difficult task, as new content-sharing websites constantly emerge. The amount of shared content makes one wonder if it is possible to battle what has already become a cultural practice of society. A whole generation is already familiar with finding the content they want, whenever they want it, without worrying about whether sharing is lawful or not. Some countries are stricter than others; in Brazil, despite it being an illegal procedure, there is no significant repression of the practice. Sharing content can be associated with freedom or, from a corporation's point of view, piracy.

Yochai Benkler (2016) wrote about degrees of freedom and power today against the backdrop of the internet's restrictions and possibilities. He draws attention to the power of the Netflix platform, whose business model is that of subscription to the content menu, which pressured the World Wide Web (W3C) to adopt *Digital Rights Management* (DRM)[1] as the standard for HTML5,[2] a way to inhibit piracy in streaming content. This way, the company now controls who can and cannot see its content. More than a question of legitimacy or legality, Benkler (2016) regards it as a matter of power through technology. For the researcher, there is a new balance of cultural power – and the end of one of the mechanisms that made the internet a place of social, economic and cultural decentralization. For Benkler (2016), we are watching the internet become, in many ways, mainly the desire for control – a mass media.

Muanis (2018) considers that the practice of *binge-watching* is not so new and gives as examples watching multiple episodes of a series from DVDs and the illegal download of episodes followed by a marathon. These practices created a new model of distribution and enjoyment of the audiovisual product, and were absorbed as a business model by platforms.

Free sharing of commercial content is not always done by people aiming at a profit. Television networks cannot and should not assume that all so-called pirates are enemies to be fought at all costs, or they will discharge part of their audience. Often the objectives of those who distribute free content owned by a TV station are quite different. At the same time, it is difficult for a corporation to create differentiated strategies on content sharing websites. A team would have to analyze the material to find out what may be a clip in honour of the main couple of a telenovela, which helps to promote the work, and what is simply a chapter shared in full. In any case, there are several strategies to circumvent mechanisms such as video fingerprinting, which can be seen in the amount of material that continues to be shared; for instance, blurring the edges of the images and enlarging it.

Graziela (2007), writing about *fansites*, notes that the logic of sharing programmes is based on donations. Fans provide links for viewing or downloading episodes, information related to characters, the universe of the series and derivatives. The more a fan contributes, the more they will be recognized. Mauss (2004) studies form and reason for exchange in archaic societies and suggests that we adopt, as a principle, giving in a free and obligatory way, giving as much as we take, always giving back. "A considerable part of our morality and our lives themselves are still generated with this same atmosphere of the gift, where obligation and liberty mingle. Fortunately, everything is still not wholly categorised in terms of buying and selling" (Mauss, 2004). This relationship between fans, mentioned by Graziela (2007), which resembles the principle of the gift, excludes producers of content and holders of the rights of the work from the negotiation.

Fans produce and share content, targeting fans like themselves. They do not necessarily want to get the attention of the television station. The dedication of fans to the fictional universe is not transferred to the producers. Graziela (2007) reveals the conflict between fans who share the programmes they love and producers who try in every way to prevent this from happening, even resorting to threats. The researcher mentions, as an example, an international television series that was negotiated for Brazil and the clash between the foreign producers and Brazilian fans who shared links of the episodes on a social network. As long as the audience and sales by windows/territories are part of the remuneration model of audiovisual content, there will be an imbalance of interests between fans and producers, causing friction.

5.2 An old or unique business model?

On 7 June 2014, *The Economist* published the article "Globo domination: Brazil's biggest media firm is flourishing with an old-fashioned business model" (The Economist, 2014). In the text, the magazine showed genuine perplexity with the "old-fashioned" business model of the station, one of the largest in the world, and especially with the audience that open TV still manages to have.

> When the football World Cup begins on June 12 in Brazil, tens of millions of Brazilians will watch the festivities on TV Globo, the country's largest broadcast network. But for Globo, it will be just another day of vast audiences. No fewer than 91m people, just under half the population, tune in to it each day: the sort of audience that, in the United States, is to be had only once a year, and only for the one network that has won the rights that year to broadcast American football's Super Bowl championship game.
>
> (The Economist, 2014)

The article states that the recording studios of TV Globo, with actors and hired crew, resemble the golden age of Hollywood, a symbol of the American film industry. Besides, it is a business based on massive ratings allied with advertising. According to the article, the fact that the telenovelas last a few months and end without a prequel or sequel is a sign of an old format of television dramaturgy. In his statement to the magazine, Roberto Irineu Marinho, entrepreneur, shareholder and then president of Grupo Globo, says that he monitors the successes and disasters of foreign broadcasters and, for him, the fact that trends and changes in the media take longer to reach Brazil is an advantage. The network's strategy is not to force the changing habits of the millions of Brazilians who still turn on their television sets in their homes and watch the block of prime time telenovelas and journalism traditionally, but to prepare for possible transformations in the market. Since 2014, the year of the article's publication, aspects of the Brazilian audiovisual scenario have changed. However, even though digital platforms are currently flourishing – Globoplay, Globo Group's digital platform, along with them – publicity in open television is still the primary business model for the telenovela.

Another way to see the current moment of popular Brazilian television would be to reflect on whether the model is old or if it is unique, and along with this, to evaluate the strength of the Brazilian telenovela, responsible for this phenomenon of persistence of a mass audience; some of the narrative characteristics of this format, historical and cultural; its transformations and recent narrative extensions.

Brazil has five private television stations and two public stations. Even though millions of Brazilians still watch network television during prime time on a television set, in 2018 the country had 78 OTT (over the top) platforms, which transmitted 139 live channels and offered a repository of 72,000 movies and 12,900 series (Lopes and Gomes, 2019). The term OTT originated during World War I and was the order given to leave the trenches and directly fight the enemy. Currently, it designates internet video services delivered directly to the consumer. Netflix, available in Brazil since 2011, is the most popular service in Brazil, with 18% of the market, followed by Globoplay, available since 2015, with 4%.

According to the Ibero-American Observatory of Television Fiction, Brazil is the eighth largest VoD market globally, but this is the main form of watching fiction for only 8% of the population. It is noteworthy to observe that according to research group Kantar Ibope, even with the cord-cutting phenomena in Brazil, cable had a 20% increase in viewing between April and July during the 2020 pandemic. Open TV's daily audience has also grown 23% in relation to the same time in 2019 (during the Soccer World Cup, very popular in Brazil). VoD digital platforms reduced the quality of transmission in Brazil since more people were using the internet during the pandemic and increased the numbers of viewers. Globoplay, the Globo Group digital platform, announced through *O Globo*, the group's newspaper, a growth of 128% in May 2020. Netflix announced 2.9 million new subscribers in Latin America in April during the pandemic.

Even with a multiplatform population, it is common for the 9PM telenovela to reach 35 audience points. In 2020, each point is equivalent to 260,558 households and 703,167 viewers in the Painel Nacional de Televisão (National Television Panel; PNT), which estimates the TV audience in 15 of the main markets in Brazil. Therefore, a chapter with 35 points means an average audience of more than 24 million people in these cities. However, this does not mean that new ways of watching this content are not emerging in the country. The 2014 *Economist* article admits that the predictions about the end of mass ratings of Brazilian open television have been around for two decades without, in fact, materializing. Thus, the changes and transformations that most television networks are going through in Europe and the United States do not necessarily occur the same way in Brazil. Brazilians have not traded one medium for another but added the new platforms to their routine.

Media corporations use social networks to promote products and seek greater exchange and public participation in programming. On the one hand, there is an effort to control the reading of the content or primary texts that are amplified by social tools; on the other hand, there is experimenting and learning of a new market component that directly influences the company's business model. Catherine Johnson (2019) analyzes the audiovisual industry's control systems in the transition from native television industries to digital platforms. For Johnson (2019), there are four dimensions of control: technological infrastructure, technological devices, online television services and content for online television. The author studies how technology control can shape our access to services, how controlling services can impact device success and how content control can also impact service success. Brazilian television stations have as challenges technological infrastructure and online services.

TV Globo already shares its production through streaming on its digital platform as well as scenes of every telenovela on the air for catch-up. Telenovelas have been in the top ten most-watched content since the beginning of the Globoplay platform, with no adaptation of format or length. Telenovelas that are on the air have a larger audience than the ones from the collection. This popularity is because being on the air works as a promotion. Nevertheless, the business model of Brazilian telenovelas remains based on advertising. Ricco and Vannucci observe: "Advertising and television have established parallel lines of growth and strong relationship between one and the other" (2017).

On TV Globo, the main commercial product today is the "Telenovela III," which airs soon after *Jornal Nacional*, prime time news at around 9:30PM, with an engaged audience of more than 24 million people daily in the 15 main cities. TV Globo offers several opportunities to advertisers interested in exposing brands and products and associating them with the narrative: national commercial; local commercial policy; national line sponsorship; merchandising and brand visualization within the product; and, currently, the extension of these actions into internet content. These actions, within commercial logic, make it possible to produce this content for television and make it profitable. It is necessary to ponder, though, on which way the public understands these advertising interventions and reacts to them, and how the consumption and advertising of the telenovela inter-twine in this narrative.

5.3 The consumption of Brazilian television dramaturgy

When a telenovela is successful from the audience's point of view, it means that the advertisers' interest will be higher, and the company will profit from the product. Almeida (2001, 2002) states that by portraying various lifestyles, the telenovela educates a consumer society. Lívia Barbosa (2004) points out that every society consumes, but that "consumer society" is not the same as the "culture of consumption," one of the definitions for Western society today. The critical approach to consumption stands out, but there is also the perception that consumption can be an act of freedom, of choice. These are parameters by which identity is expressed. To accompany this reasoning, Rocha (2005) observes that consumption is central in daily life; it is a structure of values and practices that regulates social relationships, builds identities and defines cultural maps. Consumption, for him, is the exercise of a world classification system. Colin Campbell (2006), mention-ing authors who research consumption and identity, argues that identity is the reaction of consumers to the products they want, not the products themselves. In a telenovela chapter, over commercial breaks, millions of viewers are exposed to brands, practices and styles. The narrative provokes, expands and amplifies the desire for consumption, building identities and stimulating longings and desires through their plots, conflicts and charac-ters. On the other hand, if we assume that the audience is not passive and that the final reading of a text is the viewer's, each person will understand and choose what to identify with in a unique way. Even if the narrative is absorbed massively, the reception is necessarily singular, which corrobo-rates the theories that associate consumption with identification, applied here to the telenovela.

According to Campbell (2001, 2006), there is a romantic ingredient in culture that plays a crucial role in modern consumerism. Unlike traditional hedonism, related to the consumption of pleasures, modern hedonism brings a shift from the primary concern of sensations to emotions. A tel-enovela survives precisely on emotion, conflict and the identification of the audience with the characters, the plot and the universe portrayed. According to Almeida, the association between women, consumption and emotion is partly responsible for the commercial success of telenovelas, "a program thought equally as feminine, and that creates emotional identifications with its viewers" (Almeida, 2002); and, according to Campbell (2001), it is the imagery that characterizes the search for this emotion and this pleasure. The consumer imagines his satisfaction through reverie; thus, the telenovela becomes a kind of conductor or inspiration for daydreams. If the modern consumer expresses in reality what he or she already enjoys in the imagination, television and its unique relationship with the viewer only heighten this experience.

In *Totalmente Demais/Total Dreamer* (2016), for example, the character portrayed by actress Juliana Paes is Carolina Castilho, a woman of humble origins who becomes the editor of one of the chief women's magazines in the country. Clothes or accessories that the character used on television immediately became objects of desire. The actress is also one of the main stars of the station and represents several brands, ranging from beauty products to banks. The affection and admiration for the actress and her character, with strong aspirational characteristics and the public's identification, are ingredients that contribute to transforming everything that Carolina wears into a fashion trend, which promises satisfaction of the imaginary. Morin (1989) researches the relationship between movie stars and the complexity of the subjective dimension that fans build around their idols. He discusses how a star is born, that is, what system transforms a person into an idol and how the cinematic aesthetic of classic narrative cinema works towards this purpose. Telenovelas have a more naturalistic aesthetic than classic narrative cinema. However, the phenomenon studied by Morin is that of love and projection-identification with the character, the imitation of hairstyles and costumes. When searching on Twitter, in Portuguese, for "Juliana Paes," "Carolina Castilho," "telenovela," "bracelet," "clothing" or "costume," we may find some of these reactions. Here, I have reproduced and translated some of these tweets during 2016 and 2020 – the telenovela was rerun during the 2020 pandemic, with higher audience ratings than the first exhibition.

@machadoomari 31 May 2016: "For me, the best thing in Total Dreamer was Carol Castilho's style."

@carol_arj 12 April 2016: "I only wanted Carolina's wardrobe in Total Dreamer."

@guesdriRoMilena 30 May 2016: "I only wanted Carolina Castilhos's clothes."

@karollink 10 May 2016: "Everyone look at this bracelet-ring that Juliana Paes wore in the telenovela! A new trend . . ."

@nalicetuita 30 May 2020: "This bracelet that Carol wears is VERY BEAUTIFUL #TotalmenteDemais"

@TyhSantos 25 May 2020: "I drool all over Carol's clothes #Totalmente Demais"

@AndreaRocha36 4 June 2020: "Guys, Ju Paes looks beautiful with that hair! I am very inspired by the style of the character Carol, from Total Dreamer! A delight to see this telenovela again!"

According to Campbell (2001), reality can never provide absolute fulfilment of daydreams, because wearing the bracelet or the character's clothing
will never transform the viewer into Carolina Castilho or the actress playing
her; each purchase leads to disappointment and the need is extinguished.
What is not extinguished is the fundamental longing that the reverie gener-
ates. The tension between illusion and reality creates longing as a perma-
nent habit. As observed, the telenovela, besides being a conduit for fantasy,
can amplify existing daydreams. Juliana Paes' character's bracelet will give
way to another product.

TV Globo did not sell directly any of the characters' clothes or jewel-
lery, though the viewer could easily find them on the internet and in the
many shops advertising the products. Nevertheless, publicity through merchandising is possible. Lopes and Gomes (2019) investigate the unprecedented merchandising in *Segundo Sol/A Second Chance* (2018). Through
a partnership between the station and a retail network of furniture and appli-
ances, the telenovela offered the possibility of buying furniture and utensils
that appeared in the show. The merchandizing project also had actions that
involved the morning show *Mais Você*, the exhibition of the scenarios in
subway and train stations in São Paulo, the use of the telenovela brand on
digital platforms and outlets, as well as a television campaign.

Merchandising actions of products are common in a telenovela. Actions
are generally not discreet; on the contrary, television has a pleonastic style,
and merchandising actions follow this nature. The viewer does not reject an
action like this, even when the narrative ceases to flow naturally to meet the
demands of advertising action. There is no drop in the minute-by-minute
ratings during a merchandising action. The viewer is already used to this.

In the case of *Totalmente Demais/Total Dreamer*, TV Globo's 7PM telenovela, the cosmetics company Avon had merchandising actions. Many
actresses who participated in the plot had contracts with competing beauty
product companies, preventing their participation. Therefore, one of the
merchandising actions ended up being performed by actor Humberto Martins, who played Germano, the owner of a fictional brand of cosmetics. The
scene was proposed by the commercial team, approved by the client and
adapted to the plot by the authors. It involved the character putting on Avon
mascara. An employee of the factory where Germano worked entered his
office when he applied the product to his eyes. Germano explained that, as
president of fictitious Bastille, a cosmetic company, it was his duty to try the
competitor's product to see if it was any good. The action was highly successful, and the company asked for similar actions with the character with
another product – lipstick.

The pact between the consumer audience of television and the work
includes the understanding that the business model of television contains

Figure 5.1 Merchandizing scene: Germano, played by Humberto Martins, tries mascara.

Source: Globo.

the viewing of commercials inside and outside the plot. This model does not distance the viewer from the main product he or she consumes and desires when he or she watches a telenovela: a good story stimulates the imagination and desires of the active television audience. At the same time, there is an expectation that part of this audience fulfils the implicit conditions of the pact between open television, advertisers and viewers, consuming not only narratives but also products.

5.4 New models still in transformation

Caldwell (2004) points out that major American television networks learned from mistakes when they ignored the arrival of cable television. The loss of audience and exclusivity, according to the author, made North American networks strategically invest in research of the possibilities of the internet and nonlinear programming. As Scolari (2009) observes, media convergence, besides being a cultural and technological economic phenomenon, admits epistemological conjunction. Thus, the big television networks learn and adapt to the new paradigm. Even though the audience rates of open television in Brazil are impressive, between 2007 and 2014, according to the Brazilian Audiovisual Agency (Ancine), open television went from a 63.07% share in the audiovisual market to 41.5% (Oca Ancine, 2016).

Large communication groups in Brazil that have dominated broadcasting channels now invest in format research, new transmission and broadcasting technologies, differentiated content and new forms of relationship with the public. Besides, they look for other business models that enable gains across multiple screens and platforms.

In recent years, all major TV networks in the world have launched digital platforms that offer OTT content. When networks initiate their digital platforms, they also adjust their strategy to protect their business model, which depends on advertising and attracting advertisers to more than one display window. Not so long ago, cable and satellite television operators were significant threats to the audience of open television, before the cord-cutting phenomenon. These businesses also begin to seek subscribers for their stand-alone digital platforms and already have viewers used to a business model that mixes subscription and advertising. The frontier between native companies of the digital age and television stations tends to get closer when it comes to searching for subscribers, video consumption and advertising.

TV Globo's digital video platform, Globoplay, was launched on 3 November 2015, along with "chapter zero" of *Totalmente Demais/Total Dreamer* (TV Globo, 2015). Chapter zero was a prologue to the first chapter aired on open television and had approximately 800,000 views at its launch. Ana Bueno, internet director for entertainment at TV Globo, was the one who suggested the initiative that was embraced by authors, actors and crew. Despite being a product linked to a large commercial company, the team's choice to participate or not in the "chapter zero" was personal. There was no practice model or contractual requirement for the creative team's participation in the extended contents of the telenovela at the time. Despite the success and the positive media response, chapter zero ended up working as a promotional piece. There was concern about not disturbing the end of the other telenovela still on the air, *I Love Paraisópolis* (2015). Therefore, chapter zero aired only after the last chapter of the previous telenovela, first on the internet and then during the weekend before the first chapter on Monday.

Currently, the actors and crew of a telenovela produce backstage coverage for the social networks of TV Globo and the website. The requirement not to show anything behind the scenes was removed in 2017. There is no monetization of this material, but what was a practice of actors and staff inhibited by the station became, in a short time, promotional material. Most of the chief shows and broadcasters in Brazil already have fan pages and promotional videos on YouTube and Facebook, and profiles on Twitter, Instagram and other social networks. The public interacts with these profiles and forums, participates and, mainly, shares their opinions and content production. It is necessary, however, to observe a difference between registering what is happening in the studio and creating and producing a unique transmedia narrative. But, despite all the ongoing experiences in television, such as transmedia content, spin-offs and productions on digital platforms, large corporations will focus on initiatives that are financially profitable or effectively promote valuable content.

To date, there are four major known models of monetization of audiovisual products in any media, platform or screen: publicity, in its various forms; syndication, or selling the work to other platforms and distribution/ display companies; the licensing of products which are part of the narrative universe; and direct sale of the product to the consumer, either by admission, subscription of a content menu or on-demand. I reckon that a fifth model of monetization associated with audiovisual content in digital platforms has been growing recently: content elaborated to attract users for the construction and possibly the subsequent sale of a database.

Licensing merchandise is a strategy that many television series with annual seasons already pursue. In Brazil, it is not new. The children's cable channel Gloob, part of Grupo Globo, for example, already offers several products linked to their long-lasting series. One of the difficulties of exploring the licensing of telenovela products is precisely its format. The merchandise exploitation window has to match the exhibition, which averages five to six months in length. Then, all the station's efforts will go into the promotion of the next telenovela.

Syndication, one of the examples of monetization of videos, is not a novelty for networks either, much less for Brazilian telenovelas to be exported worldwide. However, many broadcasters sold content to digital OTTs, such as Netflix, without predicting that they would be competing directly with their future digital platforms, dividing the audience. Also, as the convergence of screens and media becomes a reality, it is necessary to protect territory in every media. *Totalmente Demais/Total Dreamer* has been licensed to over 100 countries, including the United States, India, Israel, the Middle East, Germany, Georgia, Uruguay, Mexico and Chile both for open and cable television stations. The subscription business model must protect the territories of these sales.

Netflix has produced original content in more than 17 countries, including Germany, India, Japan, Mexico, the United Kingdom and Brazil. Catherine Johnson (2019), studying the effect of global digital platforms such as Netflix and Amazon, notes that there is a dimension of horizontalization of content. In the search for transactional works that appeal in more than one country, these contents effectively lose the cultural singularity of the territory. However, diverse content offer is only one dimension of the entrance of transnational digital native companies in Brazil. By October 2020, the activities of Netflix and other OTT video digital platforms had not yet been regulated in the country. Silva (2018) remarks that the challenges for VoD regulation in Brazil are to observe existing tax laws, ensure isonomic treatment concerning other audiovisual services and regulate extra-territorial offers and content sharing platforms. However, new practices of spectatorship and new business models advance faster than governments.

Globo Group uses various strategies to attract subscribers to Globoplay as a new form of monetization for Brazilian content produced by the company. The series *Justiça/Justice*, for example, was exhibited on the television station on 22 August 2016. However, subscribers of the digital platform could see the first chapters online before they aired on open television. In September 2016, the series *Super Max* had all of its chapters, except the last one, available online to subscribers before being exhibited on TV Globo. This strategy becomes apparent with series such as *Carceireiros/Jailers*, which won the Grand Jury Award at MipDrama Screenings in Cannes in 2017. Globoplay offered the whole series for its subscribers even before having a premiere date on open television. American broadcast network NBC also had a similar experience with the series *Aquarius* in May 2015. The company made all episodes available on various digital platforms as soon as the first one aired on NBC, anticipating the practice of binge-watching.

One of the chief characteristics of the telenovela is that it is to be written and produced while being exhibited so that the authors and producers may tune the narrative according to dialogue with the audience. According to Eneida Nogueira, director of research at TV Globo until 2017, the public uses social networks to comment while watching the telenovela in linear programming. The audience values this practice. If all chapters are produced and made available in advance, part of what makes the telenovela such a popular product can be lost. Nevertheless, Globoplay also tested making chapters of the 6:30PM telenovelas available one day in advance for subscribers with success, maintaining audience rates on television.

From a consumption point of view, the popularity that telenovelas still have in Brazil and worldwide is unquestionable. It is already known that the advertising model, with blocks of content and advertising, merchandising and product placement within the work, is not unshakable, especially with the advertising exodus to digital platforms – not necessarily to those linked to broadcast television. As the mediatic ecosystem evolves and business models converge, the profits from the subscription of Globo Group's digital platform Globoplay may or may not pay for these productions in the future. This depends on how this content slides to new platforms and adapts to different practices in spectatorship. Up until today, it is advertising that pays for the 150 chapters of a telenovela. It remains to be seen whether the telenovela will lose its value as a social bond if it ceases to have a "live" quality and to be an open work. Globo Group has announced for

2021 the production of a sequel for the telenovela *Verdades Secretas/Hidden Truths* (2015) by Walcyr Carrasco, winner of the International Emmy for Best Telenovela in 2016. The original show had 64 chapters and was exhibited in the 11pm time slot, which has shorter telenovelas and content ratings that permit erotic and violent scenes. The plot of the original telenovela mainly features the story of Angel, played by Camila Queiroz, who becomes a prostitute lured by a modelling agency and has to deal with the obsession of one of her clients. The sequel will have 50 chapters and for the first time a telenovela will premiere on the digital platform first, since it is a Globoplay original production. This is not a typical primetime telenovela, with over 150 chapters and produced while being exhibited, but perhaps this experience, if successful, will be the beginning of a new business model for telenovelas.

In the next chapter, through four case studies, new possibilities of sliding content from broadcast television to digital platforms, transmedia production for telenovelas, and the relationship of the public with participatory content in daily television will be investigated. With these cases, we will observe the degree of permeability of the telenovela in the face of the scenario of convergence of media, screens and transformations in spectatorship.

Notes

1 DRM is a set of access control technologies to restrict the use of proprietary hardware and copyrighted works. Ms are forms of control that already existed, for example, in 1996 to protect DVDs by an encryption algorithm for content.
2 HTML is a mark-up language for the World Wide Web and is a key technology of the internet.

Bibliography

Allen, R. (1992) Audience-oriented criticism and television. In: Allen, R. (ed.) *Channels of discourse, reassembled, television and contemporary criticism*. London: Routledge.
Almeida, H. (2001) *"Muito mais Coisas": Telenovela, Consumo e Gênero*. Ph.D. Thesis, Universidade Estadual de Campinas, Instituto de Filosofia e Ciências Humanas.
Almeida, H. (2002) Melodrama Comercial: Reflexões sobre a Feminilização da Telenovela. *Cadernos Pagu*, (19), p. 7.
Barbosa, L. (2004) *Sociedade de Consumo*. Rio de Janeiro: Jorge Zahar Editor.
Benkler, Y. (2016) Degrees of Freedom, Dimensions of Power. *Daedalus*, Vol. 145 (1), pp. 18–32.
Caldwell, J. (2004) Convergence television: Aggregating form and repurposing content in the culture of conglomeration. In: Spiegel, L. and Olsson, J. (eds.) *Television after TV*. Durham: Duke University Press.

Campbell, C. (2001) *A Ética romântica e o espírito do consumismo moderno*. Rio de Janeiro: Rocco.

Campbell, C. (2006) Eu compro, logo sei que existo: as bases metafísicas do consumo moderno. In: Barbosa, L. and Campbell, C. (ed.) *Consumo, cultura e identidade*. Rio de Janeiro: Editora FGV.

The Economist. (2014) Globo Domination. [online] Available at: www.economist. com/news/business/21603472-brazils-biggest-media-firm-flourishing-old-fashioned-business-model-globo-domination. Accessed: 26 July 2020.

Graziela, L. (2007) Fansites ou o 'consumo da experiência' na mídia contemporânea. *Horizontes Antropológicos*, Vol. 13 (18).

Jardim, L. (2020) O Lucro da Globo. *O Globo*. [online] Available at: https://blogs.oglobo.globo.com/lauro-jardim/post/o-lucro-da-globo.html. Accessed: 26 July 2020.

Jenkins, H. (2006) *Convergence culture: Where old and new media collide*. New York: New York University Press.

Jenkins, H., Ford, S. and Green, J. (2013) *Spreadable media: Creating value and meaning in a networked culture*. New York: New York University Press.

Johnson, C. (2019) *Online TV*. London: Routledge.

Kogut, P. (2020) O amor dos brasileiros pela telenovela só cresce. *O Globo*. [online] Available at: https://kogut.oglobo.globo.com/noticias-da-tv/critica/noticia/2020/07/o-amor-do-brasileiro-pelas-novelas-so-cresce.html. Accessed: 26 July 2020.

Lopes, M. and Gomes, G. (2019) *Television distribution models by the internet: Actors, technologies, strategies*. Porto Alegre: Sulina.

Mauss, M. (2004) *Ensaio sobre a Dádiva*. Rio de Janeiro: Cosac Naify. p. 294.

Mittell, J. (2015) *Complex TV: The poetics of contemporary television storytelling*. New York: New York University Press.

Morin, E. (1989) *As estrelas: mito e sedução no cinema*. Rio de Janeiro: José Olympio.

Muanis, F. (2018) Entre imprecisões e retórica: em busca de uma definição mais ampla de televisão. In: Ladeira, J. (ed.) *Televisão e cinema [recurso eletrônico]: o audiovisual contemporâneo em múltiplas vertentes*. Rio de Janeiro: Folio digital, Letra e Imagem.

Nielsen Research Institute. (2015) *The Force Is With Them: The Buying Power of Star Wars Fans*. [online] Available at: www.nielsen.com/us/en/insights/news/2015/the-force-is-with-them-the-buying-power-of-star-wars-fans.html. Accessed: 21 January 2019.

Oca Ancine. (2016) Available at: http://oca.ancine.gov.br/televisao. Accessed: 26 July 2020.

Pratten, R. (2012) *Getting started in transmedia storytelling: A practical guide for beginners*. Copyrighted material.

Ricco, F. and Vannucci, A. (2017) *Biografia da Televisão Brasileira*. São Paulo: Matrix. Vol. 2, p. 447.

Rocha, E. (2005) Culpa e Prazer: imagens do consume na cultura de massa. *Comunicação, Mídia e Consumo ESPM*, Vol. 2 (3).

Scolari, C. (2009) Ecología de la hipertelevisión: complejidad narrativa, simulación y transmedialidad en la television contemporánea. In: Squirra, S. and Fechine, Y. (eds.) *Televisão digital: desafios para a comunicação.* Porto Alegre: Sulina.

Scolari, C. (2015) Narrativas Transmídias: Consumidores implícitos, mundos narrativos e branding na produção da mídia contemporânea. *Revista Parágrafo*, Vol. 1 (3). [online] Available at: http://bit.ly/2hoyoUK. Accessed: 30 April 2017.

Shirky, C. (2010) *Cognitive surplus: How technology makes consumers into collaborators.* [ebook] New York: Penguin. Accessed: 10 June 2010.

Silva, A. (2018) *"Feud–Bette and Joan" na Fox Premium, sobre o duelo entre Joan Crawford e Bette Davis em "O que aconteceu a Baby Jane".* [Twitter] 13 March 2017. Available at: https://twitter.com/aguinaldaosilva/status/8411 09448861077504. Accessed: 26 November 2018.

TV Globo. (2015) *Totalmente Demais Chapter Zero* [video]. Available at: https:// globoplay.globo.com/v/4584051/. Accessed: 25 July 2020.

Valor Econômico. (2020) *Globo investe com recursos próprios.* [online] Available at: https://valor.globo.com/empresas/noticia/2020/03/17/globo-investe-com-recursos-proprios.ghtml. Accessed: 26 July 2020.

Variety. (2018) *Alfonso Cuaron Says 'Roma' Is Better in Theaters.* [online] Available at: https://variety.com/2018/scene/news/alfonso-cuaron-seeing-roma-in-theaters-home-netflix-1203086472. Accessed: 26 July 2020.

YouTube. (2016) *Brand of the Year Keynote: Richard Dickson, Mattel–MIPTV 2016.* [online] Available at: www.youtube.com/watch?v=ozR56w_cRj8. Accessed: 26 July 2020.

6 Telenovela and the new television

In the second chapter of this book, we investigated the narrative elements of the telenovela that resisted transformations throughout its approximately 70 years of existence and the main influences that it suffered throughout its trajectory. The vocation and the need to engage with society make the audience become the most exceptional transforming agent of these works. In the third chapter, we analyzed the construction of the telenovela's narrative. We pondered on the power chain of this production, which involves the author, team, station and their relationship with the public. From the audience's point of view, in the fourth chapter, we discussed how new media and social networks amplify the repercussions of the telenovela and whether this phenomenon translates into an increase in the power of the public over the narrative today. Finally, in the fifth chapter, we addressed the consumption of the telenovela, its business model since its origin, supported by advertising, and how the audience relates to it. We also addressed the new possibilities of business models that arise with the arrival of new media, native companies of the digital age and the convergence of screens.

Brazil has a multiplatform audience that continues to massively watch telenovelas either because of the social bonds they provide, the historical and cultural origin of these works, the practice already rooted in society or by the adaptation of the telenovela to the viewer's routine. However, the business model of television networks and the format of the telenovela have not yet fully adapted to the contemporary convergence of screens and narratives in the context of the Brazilian reality.

The sliding of the narrative of the telenovela to other screens and media, following the new spectatorship, needs to maintain the business model supported by advertising and, at the same time, leverage a new commercial model supported by subscriptions, as analyzed in the previous chapter. However, to this day, Globoplay has not invested in traditional primetime telenovelas, though they have significant popularity in the catalogue. In March 2020, before quarantining in Brazil (due to the COVID pandemic), telenovelas were the third most-watched content on Globoplay, following the reality show *Big Brother Brazil* (BBB) and live streaming. These were the telenovelas that also aired in linear programming. Telenovelas from TV Globo's collection were not as popular, even though some of them were still in the top 20 most-viewed content. During the pandemic, in May 2020, TV Globo had to rerun old telenovelas. *Totalmente Demais/Total Dreamer*, which I co-wrote with Paulo Halm, originally exhibited in 2015–2016, went back on the air with high ratings. Because of this, it became the fourth most watched content in Globoplay, following live streaming, the series *The Good Doctor* and assorted films. In October 2020, *Totalmente Demais/ Total Dreamer*, with its final chapters being broadcast, was the second-most watched content on Globoplay, followed by two other telenovelas that were also on the air. Therefore, it is possible to surmise that the popularity of a telenovela in linear programming directly influences the popularity of the same content on other media. Whether as catch-up viewing or because the telenovela on open television promotes the content online, the convergence of media attracts a multiplatform audience; they work well together.

In this chapter, we will examine empirical experiences that involve the main points already discussed in this book: format, narrative, transmedia narratives and sliding of the content of the telenovela to digital platforms; the relationship of the telenovela with the public and new forms of spectatorship and participation; and challenges and possibilities of new business models for the telenovela. Initially, we analyze the first experience of transmedia narrative content produced for a 9PM telenovela, the main product of Brazilian networks today; then an experiment with fanfic and the telenovela; next, a spin-off of a telenovela for the internet; and, finally, the sliding of a 9PM telenovela to the Globoplay digital platform.

6.1 The experience of the late transmedia narrative: *Passione*

Passione (2010) was a telenovela written by Silvio de Abreu for the 9PM slot. Bete Gouveia, played by Fernanda Montenegro, was pregnant when she met Eugênio, portrayed by Mauro Mendonça. Even so, the young man fell in love with her. Beth always believed that her son died in childbirth. She will discover the truth – that her son is indeed alive – when Eugênio, on his deathbed, asks for forgiveness; this is the cliffhanger of the first chapter and the basis for the first transmedia narrative of a 9PM telenovela.

At the invitation of one of the directors, Luiz Henrique Rios, I created transmedia narrative content for *Passione*. As early as 2010, TV Globo was investing in content extension experiences without a plan for immediate financial return. At the same time, the first portal dedicated to TV Globo's entertainment content, Gshow, was being implemented under the supervision of executive Ana Bueno and needed content and new proposals.

The first transmedia producer hired by TV Globo was Rafael Miranda, now one of the company's directors. He was responsible for continuing the experiences we implemented during this first month. From this experience onwards, the process of creating transmedia narratives changed: a transmedia producer joined the telenovela team from the beginning. This lasted for approximately ten years, as will be seen shortly. The main difference between the scope of work of the transmedia producer and other professionals linked to the internet and social media of a telenovela was that the transmedia producer could create fictional content from the universe of the narrative. In contrast, the other professionals mainly produced the making of the production, news connected to themes of the telenovela and interviews.

Our first proposal was to extend at least one scene of each chapter onto the internet. The strongest scene in a chapter is the cliffhanger. Therefore, after finishing this scene, one of the actors looked directly at the camera and spoke what the character was thinking. The first extended scene was precisely the one in which Bete Gouveia discovers that her son is alive in the first chapter. The choice to start with this scene was strategic. The extended content for the internet had to be shot with the original scene that would air on television. In order to do this, the commitment of the crew as well as the actors was vital. As it was the first transmedia narrative experience involving actors in the 9 P M telenovela, none of this was detailed in the contract, nor was there a production practice in place. The directors agreed to fit the extra scenes into the filming plan; however, they requested that the scenes always be within the eight hours of studio time, to have greater agility. Additionally, the actress who interpreted Bete Gouveia, Fernanda Montenegro, with a recognized and prestigious career, immediately adhered to the experience, making all other actors interested in participating. Silvio de Abreu, the author of the telenovela, not only authorized the proposal but also supervised the texts.

In addition to this initiative, three other experiences of transmedia narratives deserve attention in this telenovela for its novelty at the time. The first is the profile of some characters on social networks. The person responsible for the posts was Rafael Miranda, who, in addition to creating the extended scenes after my departure, also took care of all other transmedia actions of the telenovela.

One of the main problems that Rafael Miranda faced was that the unofficial profiles of the characters in social media – and there were many – had greater freedom. While fan profiles could publish false information, profanity and

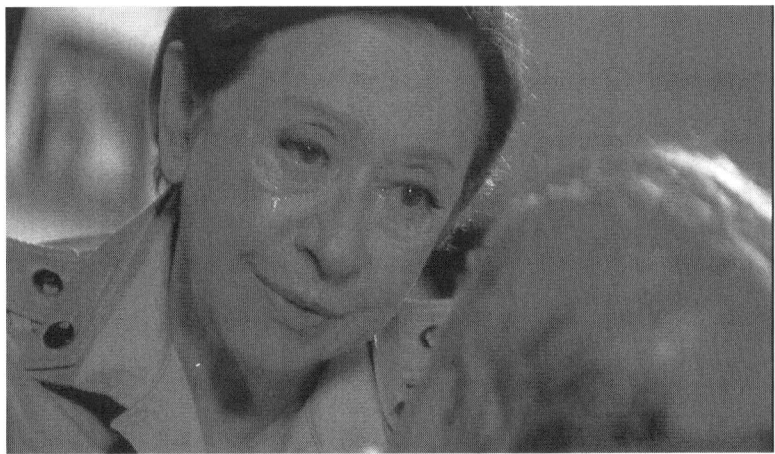

Figure 6.1 Bete Gouvea, played by Fernanda Montenegro, in the first episode cliff-hanger scene.

Source: Gshow.

Figure 6.2 Bete Gouvea, played by Fernanda Montenegro, in the transmedia narrative scene.

Source: *Gshow.*

attack other profiles, thus making them somehow popular (and fun), Rafael Miranda was forced to keep profiles consistent with the characters without revealing spoilers of the plot to the public. Consequently, the profiles he delivered were less attractive and had fewer followers than the "fake" profiles of the same characters. Borelli (2011) observed that without the movement and dynamics inherent to social networks, these official profiles contributed little to promoting the discussions around the telenovela. This role ended up being performed more effectively by the communities created by the internet viewers themselves. Rafael Miranda, representing the author of the telenovela and, consequently, corporate television on social networks, could not have the same control over the narrative that exists on television on a platform that is built by participation through social tools. Social networks are not platforms controlled by TV Globo.

During *Totalmente Demais/Total Dreamer* (2016), which I co-wrote, I personally contacted some of the fans on Twitter behind character's profiles. They were surprised, as they wrote to other fans with no expectation that someone from the telenovela would ever contact them and were happy to talk about their experience. They all met through social networks, as fans of telenovelas in general, and some because of this specific one. They consider the character profile a hobby but take it very seriously, mixing dialogues from the telenovela with their own. They see themselves interpreting the characters, like actors, and as such feel ready to improvise too. These profiles comment on the telenovela using the hashtag #TotalmenteDemais and interact with one another. One of them (@JojoDeAlcantara) said she started creating character profiles on social networks as a hobby for the series *Bones* (Fox). According to her, the official profiles would not answer simple questions such as "what is your favourite book or food?" or interact "as humans" with fans, they would only answer questions that had to do with the main narrative and nothing else. She then chose a character she related with and started talking like him on a fan board that later was taken down by the station. The group then found another platform in order to produce content: www.tapatalk.com/groups/bonesology/.

Shirky (2009, 2010), while answering how fans have time to produce so much content, writes about affection and how collaborating becomes a way for them to feel part of the work. Rafael Miranda could not compete with the number of fans acting in the native environment of these viewers: social networks. Fechine and Figueirôa (2015) consider that these viewers-producers work willingly and spontaneously for TV Globo. "That is, the one for whom the product is intended becomes a producer himself, blurring the boundaries of the old chain between producer/consumer" (p. 353).

Another transmedia experience of the telenovela *Passione* involved a character's blog related to fashion. The initiative aimed to associate the blog with a company, seeking a new business model for the internet. Although the C&A clothing chain participated in specific actions related to the telenovela, the blog did not obtain the company's adherence as an isolated product. The aesthetics of the blog, coordinated in partnership with the telenovela's costume team, followed the concept of blogs and vlogs linked to fashion and behaviour, already on the rise on the internet at the time.

Passione's last transmedia narrative initiative that deserves to be mentioned happened in the last weeks of the telenovela. A mysterious murderer started killing characters, giving the narrative a suspenseful quality. A game was proposed as an extension of the plot on the internet, and the author of the telenovela, Silvio de Abreu, became the game's main character. Throughout the week, Silvio de Abreu supplied clues on the website. Besides, the website had photos of 12 characters who were possible future victims of the mysterious killer. Each week the author "saved" a character, who would no longer be killed in the plot. The game ended with the last chapter.

In analyzing all of *Passione*'s transmedia actions, Borelli (2011) concluded that, from this experience, TV Globo's telenovelas would offer viewers new places and ways to watch television, despite observing a "conflicting articulation between models – 'old' and 'new' media – within TV Globo" (Borelli, 2011). In the author's opinion, this conflict does not assume the substitution/exclusion of one pattern for the other; they may even be complementary. However, contrary to what the author assumed would occur, there was no continuity in the transmedia experiences of the station, concerning both the volume and characteristics of the experiences, with an interruption of learning that comes from epistemological experience. In 2017 there were eight transmedia producers for telenovelas and a hundred professionals dedicated to the internet content of the station's entertainment products. However, in that same year, the transmedia producers and part of the professionals connected to social media and content for the internet were relocated or dismissed.

It is possible to assume that the budget cut was caused by the lack of financial return of these initiatives. However, these productions add substance to the work, enrich the viewer's experience and operate towards the convergence of media. These transmedia narratives already anticipated the sliding of telenovelas to digital platforms.

In television networks, even taking into account pioneering initiatives, the business model prevails. As already noted, Mittell (2015) considers that the production of a transmedia narrative may not be interesting for a broadcaster since most viewers may prefer to consume only the primary text. However, this does not mean that transmedia narrative content does not occur when promoting a telenovela, because the plot demands it or because there is an opportunity for financial return, as will be seen later.

Another critical point that *Passione*'s transmedia actions have in common is the commitment to control viewer participation. Fechine and Figueirôa (2015) argue that, because they are free individually or collectively, fans producing content have several possibilities of conduct before them: reactions and behaviours. Due to this freedom, power cannot act directly on them, even though the station may try to influence their conduct and to be involved in their actions.

6.2 A fanfic storms a telenovela: *Malhação/Young Hearts*

Malhação/Young Hearts is a TV Globo telenovela watched daily by approximately 15 million people in Brazil. Although every season the narrative approaches young viewers, the majority of the audience of this telenovela is composed of women over 35. The telenovela today begins around 6PM, which is a time for sharing between teenagers and mothers who are already in the house for various reasons. *Malhação/Young Hearts* has been exhibited from Monday to Friday for more than two decades – from 1995 to the present. Each season the author, the cast, the universe and the story changes, a new name is added to the title "*Malhação*," new sets are built and new management and production teams are hired. Currently, each season lasts a year; that is why this product is considered a telenovela for the station.

Malhação/Young Hearts is a pioneer of transmedia narrative in Brazilian television and has an extremely participatory audience. In June 2020, the show's official Facebook profile had more than 11.3 million followers and on Twitter there were over 827,000 followers on the official profile; besides, dozens of fan-led profiles can be found on all social media.

As an author for two seasons of *Malhação/Young Hearts* (2012/13–2014/15), I personally took part in practical transmedia experiences in the show; two of these were Digital Emmy nominees. The one I will analyze in this chapter used material written by fans based on the universe of the show: fanfic.

Fanfics are texts written by fans based on the universe of a book, film or television programme; they are texts derived from a work and its original universe. Souza et al. (2019) analyze fan fiction specifically in telenovelas:

> It is evident, therefore, that fanfics are written and read with the intention of broadening the audience's experiences initiated by telenovelas, through stories that focus on the viewers' adored couples, their main social and familial relationships, and the places they live in and go to.
>
> Souza et al. (2019)

Specific social networks for fanfics bring together hundreds of productions about the universe and the characters of different seasons of *Malhação/Young Hearts*. These texts usually come with a warning stating that the universe and the characters have not been created by the author of the fanfic.

Instead, the author provides a *storyline* and informs the indicative classification of the fanfic, warning viewers/readers about sensitive content such as drugs or sex scenes in their story. Fanfics can be short or have several long chapters. The vast majority are open for comments and sharing. The authors promote their fanfics on specialized social media and in several other interactive web environments.

The 2014–2015 *Malhação/Young Hearts* season was based on William Shakespeare's *The Taming of the Shrew*. We used this text to reverse the preconceived stereotype of the gender. Pedro (Petruchio) was a weak and sensitive teenager, while Karina (Catarina) was portrayed as a violence-prone athlete. Grijó and Araújo (2016) analyzed fanfics derived particularly from the universe of this season on the social network Nyah!, one of the largest repositories of this type of production on the internet. They found 115 complete texts. The shortest had a single chapter, the longest 65. All fanfics, except for one, preferred not to create new characters and used the protagonists of the telenovela. As Grijó and Araújo (2016) point out, although fanfics create their narratives, the fan still connects to the original story, creating new stories around the main romantic couples. Even so, many texts insert the characters or main couples in a different environment, for example in college, boarding schools or even in imaginary prisons. In several narratives, fans want to defend and support their favourite couple when the story seems to go the other way on the air, since love triangles are a common ingredient in a telenovela and rarely does a couple stay together from the beginning to the end of the plot.

In 2016, the fanfic "I'm Forever Yours," with 56 chapters, written by ColorwoodGirl, had 1,286 comments and the highest number of recommendations: 41. By joining Nyah! one gains access to the trajectory of this fan fiction author. ColorwoodGirl has written another 20 fanfics, most in the universe of different seasons of *Malhação/Young Hearts*, but also of other telenovelas, as well as Harry Potter and *Glee*. In the author's description, a summarized profile: "Writer. Video editor. Cover editor. Future journalist. Shipper. Boring. Maybe a little talented. But without a doubt, especially I'm Batman." When ColorwoodGirl describes herself as "writer" and "future journalist" it is a professional description, despite her using a fictitious name. Even though the content in Nyah! is amateur, Jenkins (2006) examines the commercial potential of specific fan productions. However, despite the examples of transmedia narratives that had a commercial return or yielded professional opportunities for their authors, these examples are exceptions. Individuals behind these productions do not seek market insertion, and this is not the primary vocation of these content sharing platforms.

As noted, the authors of fanfics often write to other fans like themselves and do not expect professionals involved in the production to read them. However, one of the fans contacted me on a social media platform and asked me to read her fanfic, which I did. The fan who contacted me, nick-named Anammack – with whom I still correspond with today – disagreed with the way the story was unfolding and wanted to advocate for an alternative ending to a love triangle. Through her, the entire team of writers and I were introduced to social media fanfics. We were very impressed not only with the quantity and extent of some texts, but also with their quality. As a team, we decided that some kind of dialogue between the original narrative and this material would be interesting for the show. At the time every TV Globo telenovela had a transmedia producer connected to the team. Amanda Jordão – in charge of this for *Malhação – Sonhos/Young Hearts – Dreams –* was tasked with finding a way for this exchange to happen. She came up with the suggestion of a contest that would choose a fanfic to be produced by the telenovela crew. The original idea – which was named *Malhação Fanfic/Young Hearts Fanfic –* was a collaborative experience in which fans sent their texts and their ideas from the expanded universe of the telenovela. The production crew would then shoot the idea chosen by production, script and internet professionals to air on the official *Malhação/Young Hearts* website.

Fechine and Figueirôa (2015) contend that the station, TV Globo in this case, tries to control the collaborative space of an experiment. Every call to action involves calculated risks. However, there is always the possibility of the unexpected. Bringing content produced by fans to a television programme was a risky proposal, mainly because of the copyrights involved. Copyright became the biggest obstacle for a green light from TV Globo's legal department. After six months of unsuccessful negotiations with different departments and executives, there was already a risk that the end of the season would arrive before the experiment even took place. As already pointed out, in all seasons, the author and the creative crew changes along with the cast, so there would be no guarantee that the professionals on the following season would be interested in the contest *Malhação Fanfic/Young Hearts Fanfic.*

A broadcaster as large as TV Globo operates powers of resistance and control. So if most people in the production were willing to try something new, even if that meant more work, many executives still considered it an excessive risk. In many ways and in different cases, the balance between resistance and control arises in all new initiatives in a large company. Despite the risk, the project was considered innovative and thus carried out.

All fans who participated were informed of the conditions of the contest on the programme's website. They needed to forego the copyrights of their texts to TV Globo and agree to the regulations. Also, the number of characters in the fanfic scene was limited to one page or 2,048 characters.

The legal department reasoned that accusations of plagiarism would depend on concrete evidence by the one who claimed it. In this case, TV Globo is the exclusive copyright holder of the original work, including elements and characters, and allowed fans to use these rights only for the fanfic contest. Thus, it was possible to carry out this project. As Fechine and Figueirôa (2015) observe, the relationship between prudence and risk with the new configurations of a global transmedia project is not static. The fanfics were then posted on a blog and linked to the programme's website. The project had numerous repercussions, amplified by social media.

The first contest received 4,801 entries, and the selected story, "Bianca and Her Two Husbands," aired not only on the site but also in a chapter on television. The selected fanfic mixed the universe of *Malhação/Young Hearts* with that of a famous romance by Jorge Amado: *Dona Flor and Her Two Husbands*, which deals with a love triangle between a woman, her current husband and the ghost of her ex-husband. The winner of the first fanfic initiative (there were two) was a 15-year-old student from Rio de Janeiro, Ana Carolina Souza, who had the opportunity to see her scene recorded in the studio. The author of the second fanfic, Thamires Santos, was also a teenager and a student, this time from Porto Alegre, in the southern part of Brazil. She wrote a terror story involving the protagonist, Pedro, and his father-in-law. We adapted her text to a nightmare and also had it aired on television.

Figure 6.3 Winner of the fanfic contest Ana Carolina de Souza (right) with the actor Arthur Aguiar.

Source: Globo | Isabella Pinheiro | Gshow.

Figure 6.4 Winner of the fanfic contest Thamyres Santos (right) with the actress Bianca Hamú.

Source: Globo | Isabella Pinheiro | Gshow.

The authors of the telenovela were responsible for the adaptation of the fanfic for television. There was no substantial change of the content, only an adaptation from prose to script format.

The experience with the fanfic project was positive despite the risks, but the increasingly hazy boundary between amateur and professional content may trigger conflicts. What is the difference between a professional author and an amateur fanfic author, since both create scenes that air on television? Once fanfic leaves the sharing environment of social media such as Nyah!, would amateur productions change categories?

The scenes written by Ana Carolina de Souza and Thamires Santos, winners of the fanfic contests, were written with the understanding that they would be amateur content for a contest bound by rules and promotional characteristics. These particularities of the production are an integral part of this content and permeate the trajectory in different media. Besides, both winners wrote and won the contest, but that does not mean that they are capable or even want to write a telenovela. Writing a telenovela requires not only good ideas but also the techniques necessary to develop them in a narrative with specific characteristics, as demonstrated in Chapters 1 and 2 of this book.

Jenkins, Ford and Green (2013) write about the volunteer work of fans channelled into capitalist practices. It is difficult to establish a balance of value when very different currencies are part of the exchange: financial return, recognition, affection, prestige. Moreover, this type of relationship has already become a practice in hybrid productions involving professional and amateur content, carried out through collaborative platforms and social tools. It is undeniable that the fanfic experience added value to the programme, and promoted and disseminated *Malhação/Young Hearts*. However, what did the fans get in return? In this type of relationship the recognition of the people who produce the programme was a bargaining chip for Ana Carolina and Thamires, as was visiting the set, meeting the actors and the production team. It is a statement that original work produced by a fan also has quality and value. Nevertheless, we have to consider: was this a fair exchange? Did TV Globo gain more from the experience or did Ana Carolina and Thamires?

Jenkins, Ito and Boyd (2016), revisiting research on the culture of participation of 2004, ask whether it is possible to have a relevant, significant exchange in a situation controlled by a corporation. The ability to create and share content is separate from the ability to govern the platform where the content circulates. For these authors, participation could become exploitation. It is necessary to highlight that all major content-sharing platforms are properties of private companies. These companies archive personal and collective data, stories and productions on private servers for commercial purposes and use this information and content without, necessarily, the authorization of users.

This boundary between voluntary work and exploitation, between recognition and questionable use of creation, amateur and professional, is yet to be negotiated and understood in the face of the new mediatic ecosystem. Most likely, this is a relationship that will never be crystallized or perfectly balanced. Jenkins et al. (2013) claim that the status of what is exchanged online is hybrid, and often what has sentimental value for one party has commercial value for another. The experience of the fanfic project is one such case. Nevertheless, once the pact is established and the balance of expectations achieved, albeit with different objectives, there may be a point of balance between a major television station and teenage fans. Benkler (2016) states that, unlike early generations, companies today are very aware of the importance and value of networking activities. However, the way companies use this knowledge to their advantage can also change and even destroy the trading experience. Each experience and collaborative initiative, involving corporate power or not, should be evaluated from an ethical point of view. In the case analyzed in this chapter, for both sides the end was satisfactory and with measurable gains. After all, a possible parameter is precisely the balance between expectation and reality.

The fanfic project was just one of several initiatives that television can use today to learn and reinvent itself. Television is made up of people and, as much as they are inserted into a large corporation, they also use social media to navigate and relate to content they admire. The experience with fanfics, as well as others of the genre, showed that television fiction is permeable to collaborative content, even if it is in isolated initiatives. The fanfic contest did not continue. As pointed out in the previous chapter, an initiative that does not prove financially profitable is likely to be interrupted.

6.3 The transmedia narrative as a business: *Totalmente Demais/Total Dreamer* spin-off

In addition to other initiatives of transmedia narratives related to *Totalmente Demais/Total Dreamer* (2016), one month before the end of the telenovela, the station executives suggested the production of ten spin-off episodes for the internet after the telenovela ended on the air. Spin-offs of telenovela plots and characters are not new. The series *O Bem Amado/The Well-Loved*, exhibited between 1980 and 1984 by TV Globo, originated from the homonymous telenovela written by Dias Gomes in 1973. Mário Fofoca, a clumsy detective, played by Luís Gustavo and created by Cassiano Gabus Mendes for the telenovela *Elas por Elas* (1982), became a television series the following year and afterwards a film. One difference between these experiences and of the *Totalmente Demais/Total Dreamer* spin-off is the media: the internet, with its characteristics, such as nonlinear display, allows the viewer to choose when, where and how to watch.

The experience of the *Totalmente Demais/Total Dreamer* spin-off was also an opportunity to observe if the extended content of a successful telenovela on the internet would harm the audience of the new one on the air. The strategy of the company up to then was to avoid the extension of the telenovela and, consequently, of affection.

After a meeting with the team, we agreed that the series would be an opportunity for collaborators to get credit as writers for each of the episodes they would be entirely responsible for. Collaborators or staff writers usually write only dialogues of scenes already described in step outlines by the telenovela's head authors. Moreover, the work of the collaborators would decrease in the final stretch of the telenovela, which was not the case for the head writers – we proposed only to supervise the spin-off.

The writing team conceived the episodes according to a previous agreement with the production and directors of the telenovela, using fewer actors and compact sets. As observed earlier, a telenovela is written with several different plots and nuclei to make the product viable. For the spin-off, we chose a nucleus that had a humorous streak. We assumed that it would be more compatible with the media – the internet – and this nucleus also had actors who had greater visibility on social networks. The company adjusted salaries, extending the contracts of those involved until the end of the spin-off production.

Unlike the chapter zero experiment, the spin-off had no promotional value for the telenovela, since the mothership narrative was no longer on the air. Therefore it had to be profitable on its own.

Avon had exclusive sponsorship and merchandising rights in *Totalmente Demais/Total Dreamer*. The spin-off was an opportunity for the station's commercial department to seek new partners among beauty products companies, since Avon showed no interest. Risqué wanted to promote a new line of nail varnish and sponsored the spin-off, with merchandising scenes in some episodes. The brand's director, Regiane Bueno, in an interview at the time remarked:

> Risqué is a modern and bold brand, hence the decision to participate in a pioneering project like this. We understand that it is important to expand the range of communication with our consumers, opening a new channel to inform about our news firsthand, including for customers who are at the same time tuned to the Internet and TV.
>
> (Adnews, 2016)

Episodes were made available on Globoplay twice a week on specific days and the total views exceeded 4 million. During the exhibition on the internet, the chapters of the telenovela on the air, *Haja Coração/Have a Heart*, had a considerable audience on television. This way, the fear of one telenovela sabotaging the audience of the other, even in different media, ceased to exist. The Brazilian audience is multiplatform. Experiences overlap and do not replace one another. Later the content of the spin-off also ended up on open television for two consecutive weeks during the afternoon programme *Video Show*.

Stations try to absorb advertisers who migrate to other digital platforms from their digital platforms, and the *Totalmente Demais/Total Dreamer* spin-off was just one example of those initiatives. The search for a new business model based on subscribers and allied to the sliding of telenovelas to digital platforms brings new challenges for broadcasters. For November 2020, Globoplay announced a spin-off of the telenovela *Malhação/ Young Hearts*, season 2017/18 by head writer Cao Hamburguer, exclusively for subscribers. Along with an original series of ten episodes with the characters years later, Globoplay is also producing a documentary about the process and a talk show with the actresses.

6.4 A telenovela in the database: watch it as you wish

As already observed, the narrative of a traditional telenovela is well known to the Brazilian public. As an author, I am surprised by viewers in focus groups or social media that often foresee the unfolding of some plots. It is not unusual for comments or opinions to become a source of inspiration, but, most of the time, the audience wants to see the characters for whom they have affection to be happy as soon as possible. However, a telenovela, or any melodramatic narrative, does not exist without significant conflicts. That is, there will be a happy ending, but until then, many obstacles will disrupt the lives of heroes the audience loves.

As already noted in the second chapter, the author's power over the telenovela is directly related to the production model, which requires six chapters to be produced per week. As the telenovela is an extensive and "open" work, by following what is happening on the air, the author can make adjustments to the original synopsis. The actors and directors contribute a lot to the result, thus also influencing the course of the narrative. However, new examples of interference in the original work emerge along with recent transformations in spectatorship, the relationship between the work and the audience, thanks in part to new technologies and interactive platforms.

Despite all the transformations concerning the convergence of screens and amplification of the repercussions of a telenovela on social media, three fundamental ingredients of daily television dramaturgy until recently had not changed. The first is the control of the narrative temporal flow; second, the continuity of the story by the author; and third, the control by the broadcaster of the display, flow and reruns of chapters. For decades, daily from Monday to Friday (or Saturday, depending on the product), chapters are made available to the public on television. Thus, the arc of the primary and secondary plots, the trajectory of the characters, their transformations and conflicts are offered linearly and continuously, even if not all viewers are as assiduous.

Chatman (1978), concerning story structure, ponders that there is the time of reading a text and the time of the narrative – in other words, how long a speech takes to happen as opposed to the time events take place in the story. He notes that many questions may be raised taking into account these time-related elements:

> For example, how is the story anchored to a contemporary moment? When is the beginning? How does the narrative provide information about events that have led to the state of affairs at that moment? What are the relations between the natural order of the events of the story and the order of their presentation by the discourse? And between

the duration of the discursive presentation and that of the actual story events? How are recurrent events depicted by the discourse?

(Chatman, 1978)

Kozloff (1992) assesses that temporal distortions help us discover the narrator on television. The closer the speech is to real-time, the more invisible and less intrusive this narrator or whoever tells the story will be. A live narrative would be the antithesis of a narrative with repetitions, parallel actions, flashbacks or flash-forwards and temporal elasticity when the narrator – in the case of a telenovela, the head author – is easily revealed. The events of a story, its order and duration are altered. Mittell (2015) also notes that for every narrative, time is an essential element, especially in the case of television narratives. We can consider three different temporal flows for all narratives: *time of story* – how time passes within the story; *time of discourse* – the duration and structure within a narrative; and *time of narration* – time that the plot is on the air, or the deadline set to tell the story. This same model of classification of narrative temporal flow also applies to a telenovela, and time is always a vital writing tool of the author. Until recently, the time of the flow of a telenovela, be it the time of the story, speech or narration, was not something that the viewer could control.

Globoplay was launched on 3 November 2015. On the computer, mobile phone or tablet – with expansion plans for video game consoles and connected TVs – the public can watch the programming through *simulcasting* (*streaming*) and have free access to excerpts of specific content. Subscribers can access entire programmes at any time, plus films and series purchased from other content producers, as well as access to exclusive cameras in popular reality shows such as *Big Brother Brazil* (BBB). Regarding telenovelas, Globoplay subscribers have the right to watch full chapters. Those who consume content for free are required to watch chapters through excerpts, with advertisements, on Globoplay. However, by making content available this way, the company allows the viewer to seize the *time of the narration* and *discourse* of the work for the first time in the history of television. Although the company sells subscribers the comfort of watching a chapter in full, inadvertently, by offering its content in excerpts, the broadcaster may be favouring a new practice of spectatorship.

Manovich (2015) notes that while cinema and romance favour narrative, the computer age introduces its correlate: the database. The author reports the predominance of the database form in new media and mentions, as an example, the website of a radio or television station. Watching linear programming through streaming is just one of the options of the database.

A first analysis of the number of online views of each excerpt available of the prime-time telenovela *A Força do Querer/Edge of Desire* (2017), by Glória Perez, shows that the selections have different audiences. On 19 June 2017, for example, the most viewed excerpt on the Globoplay platform had 88,140 views, while the least seen had less than half, 31,689 views. It is no surprise that the most viewed excerpts are the first and the last two, i.e. the chapter cliffhanger and its resolution in the next chapter. Cliffhangers have the purpose of stirring the viewer's curiosity so that he or she will watch the telenovela the next day.

In addition to the significant number of visualizations of the initial scene and the cliffhangers of telenovela chapter online, when analyzing the chapters of the whole week, it is also possible to see that specific plots or nuclei are more successful online than others. Manovich (2015), when researching the culture of the database and its relationship or opposition to the narrative, notes that a traditional linear narrative is one among many other possible trajectories. A telenovela has several nuclei and, depending on the story, there may be an alternation in the protagonism of different plots.

Such an example would be *A Força do Querer/Edge of Desire*. The telenovela launched on 3 April 2017, having as a main plot the story of Ritinha, who believes herself to be a mermaid, and her love triangle with urban playboy Ruy and lorry driver and eternal groom Zeca. Nevertheless, throughout the week of 19 June to 24 June, the telenovela narrative was focused on the plot of the character Bibi Perigosa. Fifty-eight excerpts were made available about this nucleus, while the nucleus of the mermaid Ritinha had 34 published excerpts. However, even though Bibi Perigosa's nucleus had all the cliffhangers, the scenes with the most views were those that dealt with the love triangle of Ritinha, which began the story.

The most viewed excerpt of the whole week on 24 June, with 134,520 views, is part of the mermaid plot and has the title "Zeca and Ruy are impressed with Jeiza and Ritinha." Naming the scenes for the viewer on digital platforms is a demand for new spectatorship on a new platform. Without this, content consumption and understanding of what is offered by the database would be confusing. By monitoring the number of views of the excerpts referring to this nucleus, it is possible to conclude that there are people who watch only this plot, building a specific trajectory in the database. It is also possible to remark that the excerpts of intense scenes, with sensual, dramatic or violent ingredients, have thousands more views than others, regardless of the nucleus. The scene of a car accident that same week, for example, of a nucleus that did not have as many online views, the one with the transgender character Ivana, obtained 86,191 views. The average of all the other scenes of this plot that week did not reach half as many views. Nevertheless, there may be spectators who watch only scenes of this last plot and scenes from other nuclei that are sexy, for example. Each viewer can create not only their own narrative, but also watch excerpts from a single chapter over days or excerpts a week at a time.

Despite the efforts of television networks to control their content on video-sharing platforms, it is not difficult to find fan-produced compilations of scenes of their favourite couples. *As confusões de Pedro e Karina em Malhação parte 1/The Misadventures of Pedro and Karina in Young Hearts Part 1* is an edition that has only scenes of the main couple from the season I co-wrote in 2014 and 2015. In June 2020, the video had 2,788,380 views. This video is not unique. It is possible to find short clips that feature many couples from telenovelas. This content needs to circumvent broadcasters' copyright protection mechanisms, such as fingerprinting. Therefore, many use stratagems such as blurring images, creating frames around scenes or using colour-changing filters to escape identification tools. There are many differences between the compilations present on video-sharing platforms and the customized viewing of specific excerpts on the Globoplay digital platform by a viewer. The compilations, made by fans, are narratives offered to other viewers as closed works. There is a curatorship in the video of Pedro and Karina. The compilation does not include all the scenes in the 275 chapters that were exhibited during the telenovela.

Allowing the viewer to assemble his or her narrative was certainly not the purpose of the station when offering excerpts from telenovelas online for free, but rather a catch-up mechanism in the video on demand (VoD) platform. Notwithstanding, it is possible that many viewers, now faced with the possibility of watching only the plot that interests them, prefer to watch the telenovela through excerpts. It is the *custom telenovela*. The database brings new distinct possibilities of spectatorship. The chapter of *A Força do Desejo/Edge of Desire* most seen in full by subscribers in the week of 19–24 June was on a Monday, with 54,710 views – that is, less than half of the most viewed excerpt for free.

By making the telenovela available in a database to be then reorganized by its consumers, the station removed an essential part of the author's power – the use and control of the flow of narrative time as a writing tool – and handed it to the public. Even if unintentional on the station's part, this opportunity possibly satisfies the demands of the part of the public that does not identify with the various plots and nuclei homogeneously. Taking into account all the transformations that the telenovela has already undergone – updating the narratives, division by nuclei, the influence of series, among others – perhaps a next change is precisely this new way of watching the content.

Currently, TV Globo's telenovelas occupy five hours of prime time every day, providing national narratives with famous actors and actresses displaying and shaping culture. In many ways, for many people in Brazil, the telenovela is synonymous with television. It continues to fulfil this function, but it cannot avoid the needs of new audiences. The audience in the country watches television in a "traditional" way, but the form with which the audience interacts with the content is increasingly sophisticated, amplified by digital platforms and social media.

Bibliography

Adnews. (2016) *Globo reforça estratégia multiplataforma com spin-off de 'Totalmente Demais'*. [online] Available at: https://adnews.com.br/globo-reforca-estrategia-multiplataforma-com-spin-de-totalmente-demais. Accessed: 26 July 2020.

Benkler, Y. (2016) Degrees of Freedom, Dimensions of Power. *Daedalus*, Vol. 145 (1).

Borelli, S. (2011) Migrações narrativas em multiplataformas: telenovelas Ti-Ti-Ti e Passione. In: Lopes, M. (ed.) *Ficção televisiva transmidiática no Brasil: plataformas, convergência, comunidades virtuais*. Porto Alegre: Sulina. p. 115.

Chatman, S. (1978) *Story and discourse: Narrative structure in fiction and film*. New York: Cornell University. p. 63.

Fechine, Y. and Figueirôa, A. (2015) Transmidiação: explorações conceituais a partir da telenovela brasileira. In: Lopes, M. (ed.) *Ficção televisiva transmidiática no Brasil: plataformas, convergência, comunidades virtuais*. Porto Alegre: Sulina. p. 353.

Grijó, W. and Araújo, G. (2016) Fanfictions e telenovela: ficção seriada televisiva na cultura participativa. *Vozes e Diálogo*, Vol. 15 (2). [online] Available at: https://siaiap32.univali.br/seer/index.php/vd/article/view/8841. Accessed: 3 February 2019.

Jenkins, H. (2006) *Convergence culture: Where old and new media collide*. New York: New York University Press.

Jenkins, H., Ford, S. and Green, J. (2013) *Spreadable media: Creating value and meaning in a networked culture*. New York: New York University Press.

Jenkins, H., Ito, M. and Boyd, D. (2016) *Participatory culture in a networked era*. Cambridge: Polity Press.

Kozloff, S. (1992) Narrative theory and television. In: Allen, R. (ed.) *Channels of discourse reassembled*. London: Routledge.

Manovich, L. (2015) Banco de dados. *Revista Eco-Pós*, Vol. 18 (1). [online] Available at: https://revistas.ufrj.br/index.php/eco_pos/article/view/2366. Accessed: 25 July 2020.

Mittell, J. (2015) *Complex TV: The poetics of contemporary television storytelling*. New York: New York University Press.

Shirky, C. (2009) *Here comes everybody: The power of organising without organizations*. London: Penguin Books.

Shirky, C. (2010) *Cognitive surplus: How technology makes consumers into collaborators.* [ebook] New York: Penguin. Accessed: 10 June 2010.

Souza, M., Lessa, R., Bianchini, M., Nolasco, H., Souza, B., Alves, G. and Araujo, J. (2019) Fic writers, adored couples and fictional worlds: Creation and re-creation of Brazilian Telenovelas in fanfictions. In: Lopes, M. (ed.) *World building in Brazilian TV fiction.* São Paulo: CETVN/ECA/USP. p. 76.

7   Conclusion

The first chapter of this book contextualized television in Brazil as a unique phenomenon with massive audiences, even with a cross-platform population. The second chapter addressed the main elements of the telenovela's narrative and how it has changed over its nearly 70 years of existence. This transformation was influenced by several agents, mainly Brazilian society, making the telenovela unique in Brazil not only for its content and format but also for the massive audience this product attracts. In line with several theorists, the chapter also considered that the telenovela, with its characteristics of melodrama and feuilleton, needs to be understood as part of Brazilian culture and as part of a tradition associated with Latin America. In the third chapter, the process of writing and producing a telenovela was examined. The characteristics of this process contribute to a chain of power in which the producers depend on the author, even though he or she may be subject to supervision or substitution. Working constantly under pressure from ratings and surveys, the author maintains his or her autonomy, especially if the telenovela is successful with the audience. The fourth chapter dealt with the relationships between authors, broadcasters and viewers. It investigated how the engagement of the public may be amplified by social media and interactive platforms. Also, the chapter discussed whether the power and influence of this audience has increased in relation to television content. The fifth chapter reviewed the business model of the telenovela, highlighting how new media, digital platforms and the convergence of screens transform and bring new possibilities, as well as challenges, for television. Finally, the sixth chapter analyzed empirical cases of sliding content from the telenovela to other media and the consequences of these experiences not only for the narrative format, but also for the business model and the power chain of a telenovela. The chapter also addressed *transmedia* narrative productions, with and without content generated through the audience participation – *crowdsourcing* – and the ethical implications of a relationship between corporate television and amateur content. In these chapters there are many clues to answering the following question: how may or may not Brazilian television transform through the telenovela, considering its massive audience, and survive the sliding of content to other screens and media, the arrival of digital and interactive platforms with a segmented audience and transformations in spectatorship?

The technology that accompanies these changes is part of the current transformations in Brazilian culture and society as a whole. As Crary (2012) noted, the observer's story cannot be reduced to technical and mechanical changes or changes in art and visual representation; technology is always concomitant or subordinate; it is part of other forces. The telenovela viewer of today, with his or her subjectivities, often consumes audiovisual content at the same time on several screens, managing perception and time in a way never before experienced.

The effort of the country's television companies to migrate to other screens and exhibition platforms meets a demand from the public, which, through several sharing platforms, already watches television content wherever and whenever they want, legally or not. Casetti and Odin (1998) evaluated that the communicative pact between the viewer and the broadcaster is not limited to a single formula. It involves rules, purpose, proposal, consent of the parties and negotiation. Technologies and practices that emerged with the strengthening of the internet as an interactive media on the rise in contemporaneity have only brought new elements into this negotiation.

Muanis (2018) writes about the idea of search and *satisfaction* for the individual who consumes content, whether this is something to be read or something to be watched. When searching for the content they want, the audience also slides between offline and online media. It is currently possible to affirm that there is no clear division between online practices and those of "real life"; after all, we have become connected individuals. In 2013, the UK government commissioned anthropologist Daniel Miller (2013) to study the future, identity and the internet. By defining identity as a variable, historical and cultural condition, one that is attributed and adopted – as opposed to a psychological state – he concluded that most people engage in a mix of communications with a multitude of online/offline identities, without a clear difference between them. Our online identity reveals and makes us aware that our offline identity was already multiple, contextualized culturally and historically. The internet does not represent another domain or social space – it has already integrated into our daily lives and is essential for our communication. We reveal ourselves as individuals through what we consume, comment, produce and share offline and online. Therefore, it is natural that the encounters, confrontations and conflicts of offline society also exist in the online world, even if magnified by the technological tools and practices characteristic of the internet.

The audience that watches a telenovela on a digital platform, on any device, and at the time that suits it, is not from a historical and cultural universe different from the audience that watches television within the programming stream on a television set. They may even be the same people, as already observed, but there are many differences in the consumption experience. Along with news, sports and reality shows, the telenovela still has significance when it is broadcast at the same time to millions of people. First, because it follows the routine of the household; second, because the fixed daily schedule of the telenovela within the programming flow reinforces the social bond it provides to the country and also creates the idea of belonging. How can these values and aspects of the telenovela slide, along with its content, to digital platforms and new media? What is it that sustains Brazilian television: the telenovela or the other way around? Both hypotheses are true.

The year 2013 was the first time that among the nominees for the Emmy awards there was a series made for OTT, *House of Cards*, produced by Netflix. In 2019, there were 14 different nominees streaming on Netflix, Hulu and Amazon Prime Video. Mirian de Icaza Sánchez (Svartman and Sánchez, 2018) believes that the telenovela will slide to other platforms and media as a demand from the cross-platform public and does not see an end to the telenovela narrative. Interestingly, after 26 years in TV Globo's quality control area, the researcher was recently hired by an international digital platform to, in her words, do what she did at TV Globo, that is, evaluate projects for Brazil.

The most exceptional transforming agent of the telenovela is precisely the public, its practices and the historical, cultural moment of society. So, the importance of social tools for the public is one of the clues of how the telenovela may be adapted to new media. These make it possible for the idea of belonging, which the telenovela offers, to remain in some way. As Shirky (2010) notes, these social tools make it easier for people from different areas, but who believe or desire the same things, to meet and form discussion and collaboration groups. While the telenovela is on the air, as with sporting events or the news, texts about the narrative are all over social media. However, if the telenovela is available on a digital platform outside the open TV programming flow, this exchange may be reduced, losing its "live" quality.

Muanis (2018) notes that the production of the channel *Porta dos Fundos*, bought in 2017 by Viacom, follows a television distribution model on the YouTube platform, with a date and time for the release of each chapter; that is, it emulates a schedule. Apple TV+, HBO Go and Netflix do the same with some of their series and movies, announcing the date and time when episodes will be available for streaming. Perhaps this model offers a possible transition between watching a telenovela within the programming flow and the experience of watching it at any time and on any device through the digital platform. The audience would have a common time frame to comment on the premiere of each chapter, which is closer to the experience of commenting on the telenovela online "in real time."

The convergence of screens and media is not about the substitution of one way of consuming telenovelas for another, but the sum of the two. This blend of experiences may also be the future of the spectatorship of the telenovela, which would prolong the current business model of broadcast television. New audience metrics, which include the same product appearing on different media, could keep the business model supported by advertising and even support a hybrid model, with subscriber-generated features and sales of commercial spaces. Traditionally, television uses GRPs (*gross rating points*) as metrics – in the case of Brazil, a sampling of the audience of the 15 major cities of the country. Focus groups also help in the evaluation of a telenovela and the rejection or acceptance of characters and plots. Digital platforms, in addition to the audience, use a set of performance metrics seeking to assess the level of viewer engagement. Since 2016, the Nielsen research institute has integrated, in addition to digital TV, mobile (mobile devices) to its offering of metrics for advertising, called *Total Ad Ratings* and *Digital Ad Ratings*. With the arrival of this technology in Brazil, the telenovela could remain viable, considering a business model in part supported by advertising, if it could attract a mass audience in combined media. This model could be added to revenue from subscriptions.

Catherine Johnson (2019) evaluates that the internet has worked for many years as an extension of television, with transmedia narratives, content about television programmes and other experiences outside the television set. In the last decade, the relationship between television and the internet has changed. The internet has become a medium where you can watch television content. The "Online TV" Johnson describes is forged by a series of services that make it easy to watch selected audiovisual content on connected infrastructure and devices. The author notes that there is a continuity between contemporary online video services and open and cable television, as well as the platforms and digital extensions of broadcasters, because of all these forms of watching television coexist as we do – offline and online.

Johnson (2019) compares companies that are native to television and audiovisual content companies that are digital natives. The first are content providers through broadcast, cable, satellite and digital TV. These companies, such as TV Globo, extend their services to online digital platforms and have a history of producing, acquiring and delivering content, as well as access to large catalogues of programmes. Those that are digital natives, in turn, offer online services and have a TV service for the entire internet eco-system. These companies, such as Netflix, Amazon and Hulu, have begun producing content and investing a large amount of resources into it. For the researcher, the great advantage that native television companies have over digital natives is the content collection and the already established emotional relationship of the public with it. I would add that in the case of Globo Group and other worldwide networks and corporations, such as Walt Disney Company which recently lauched the subscription VoD streaming service Disney+, the possibility of producing or acquiring content for more than one window or screen (broadcast television, cable television, digital platform) is also a strength, not only for promoting the content but also for diluting costs. However, Johnson (2019) states that a critical point is the regulation of VoD platforms in each country and the impact on production and distribution of content. By October 2020, VoD still had not been regulated in Brazil.

The future of the telenovela is in the hands of the viewer, and the future of Brazilian television is in the hands of the telenovela, the most popular content on Brazilian television for the last 70 years. Television critic Kogut (2020) observes:

> Anyone who likes to claim the telenovela is an endangered genre that has been swallowed up by American series should bite the tongue. This was more than proven during the pandemic. The series are wonderful, but occupy another compartment in the heart of Brazilian public. The telenovelas continue in a place of honour and that territory has only expanded in recent months.
>
> (Kogut, 2020)

Executives who run networks need to be vigilant when formulating their content slippage strategies for digital platforms concerning format, language and distribution.

This book has already approached quantitative and qualitative research and their relevance in defining programming and legitimizing the success of a telenovela. However, native digital companies in a new paradigm of producing, distributing and displaying content on digital platforms have a substantial advantage over their competitors: data. The data these companies acquire from their users on their platforms plays a central role throughout the business model. For example, data has a fundamental attribution in advertising, fuelling the entire campaign cycle; in offering content on subscription services, user data also feeds all other steps – from content selection and production to individualized suggestions. Data-driven audience segmentation is a crucial element in leveraging two major revenue models: subscriptions and advertising. Consumer data provides the means to expand audiences and create new forms of engagement, as well as new business models. In the case of digital audiovisual content platforms, they help refine offers, allowing you to understand who, what, where, when, how and why the content is being consumed, and optimize it even more.

The accumulation and use of data by various corporations reinforces the concept of the continuous learning environment of these companies, in real time, guiding the entire media experience. Netflix, for example, uses a feature attribution model for its content to boost the recommendation algorithm based on user data. It seeks to illustrate the offer of films and series on the platform in a personalized way. The production of original series is also influenced by this system. With this method, the Netflix platform does not need to focus on mass-appealing content; they can bet on content that appeals to specific audiences already identified by the service. This segmented and diverse audience makes up a massive audience of an unequal catalogue that no user has interest to in its entirety. Everything is tailored.

What polls once did for traditional television, with professionals like Eneida Nogueira and Mirian de Homero Icaza Sánchez at the head of research departments for a broadcaster like TV Globo, the platform itself can offer with a sophisticated algorithm. These organizations extract value from data in the form of predictions, analyze datasets that serve their goals, and thus create competitive advantages. Greater predictability is one of the main differentials, as it fundamentally transforms these companies that cease to operate as reactive environments managed by static and obsolete data. However, predictability does not replace the sensitivity of the author or creators and directors of an audiovisual work. If data were enough, all productions of these platforms would be successful, but what is observed is the cancellation and discontinuations of series, content adjustments similar to what occurs on corporate television.

In the case of Brazilian networks, it is argued once again that it is partly the strength of the telenovelas, with a massive audience, that keeps the business competitive even with the arrival of digital natives. With such a massive audience, it is not yet pressing (or possible) to collect qualitative data in real time and, consequently, evaluate the segmentation of the audience. Eneida Nogueira, the former research director for TV Globo, points out that the massive television audience caters to the desire for family life. The researcher compares some television shows to Hollywood family movies. "TV is a shared media. Some things are still common for people. When cable television came in, everybody said 'TV is going to be over,' but is everyone going to be left alone? People want to be together" (Svartman and Nogueira, 2018). Nogueira considers that the telenovela is part of the Brazilian routine and that, if there is no drastic change in this, it will continue to be popular. "You need to keep watching what happens to people all the time and then calibrate what you are offering to keep up. Routine gives security; it's familiar. People like to feel like they are in a place where they recognise what is going on." She also assesses that the telenovela can slide to digital platforms; the researcher believes, however, that there would need to be a transformation in the format, such as editing, so that telenovelas would be shorter on digital platforms. Shorter teleno-velas also influence the television production. A large part of the resources used in the production of the telenovela is the construction of outdoor sets and studio settings. This cost is diluted throughout many chapters so fewer episodes and, therefore, less air time would mean less publicity and these costs would not be covered, unless subscriptions also paid for the costs.

The telenovela, as well as its pillars, melodrama and feuilleton, was never static and has always adapted to its time. In the second chapter, we discussed the resilience of the feuilleton format and melodrama through-out history. The feuilleton went from the newspaper to literature and then to radio and television. Likewise, the melodrama has also adapted and transformed as it permeated different media at different historical moments. The melodrama has already travelled from theatre stages to radio and then to film and television. Telenovelas are edited and adapted when sold to other countries and when exhibited in afternoon reruns, with a different age rating classification. *Totalmente Demais/Total Dreamer* (2016) originally had 175 chapters, but in 2020 the telenovela was exhibited in Croatia with only 130. Therefore, the telenovela always follows the demand of the public and its consumption characteristics, as melodrama does too.

According to director Daniel Filho (2001), the main objective of televi-sion is to inform and please the public. For the director, the telenovela will remain because it has existed for a long time, it pleases, and will continue to exist in the new platforms of transmission of content over the internet. For him, the telenovela fulfils the social role of "gossip" – it produces a point for people to exchange opinions and share experiences that are similar to those they live in society. In addition to the importance of these tertiary discourses – analyzed by Fiske (1987) –Daniel Filho (2001) does not believe in a crisis; he sees change. The telenovela, therefore, will be trans-formed because the public wishes it so.

In writing about the cognitive revolution that 70,000 years ago gave rise to the history of man, historian Yuval Noah Harari (2015) argues that what made human beings different, the unique trait of *Homo sapiens*, was the ability to create and believe in fiction. All other animals use their communication systems to describe reality. We use our communication system to create new realities; as an example, he mentions fictions such as money and religion, which made collaboration and coexistence between human beings possible. For Harari, there is an *intersubjective* dimension in the communication network linking the consciousness of many individuals. The universe of a telenovela also has an intersubjective dimension, while it is on the air, with its characters, territories and plots being part of the reality of millions of Brazilians who consume, interact and relate with it in various ways. However, it is also a melodramatic narrative that permeates the history and repertoire of Brazil, adapting and transforming.

As long as the telenovela continues to be part of Brazilian culture, it will continue to be a mass product. Moreover, as a mass product, it can be consumed on any platform, media or screen and will continue to be a telenovela and, therefore, Brazilian television, even if it has a hybrid business model. Secondary and tertiary texts will be present in these various media and platforms as well.

The telenovela is the Brazilian equivalent to blockbuster films, world bestselling books, the music industry's millions of downloads on the internet and shows that fill stadiums around the planet. However, in all these examples, culture and the market go together. The telenovela depends on the management of corporate companies and the executives behind these corporations. They must understand that, when evaluating the business model, it is necessary to observe and preserve the engagement of the telenovela with the public – whether through research, user data, social networks or the sensibility of an author.

Bibliography

Casetti, F. and Odin, R. (1998) Da paleo à neotelevisão: uma abordagem semiopragmática. *Ciberlegenda*, (27).

Crary, J. (2012) *Técnicas do observador: visão e modernidade no século XIX*. Rio de Janeiro: Contraponto.

Filho, D. (2001) *O circo eletrônico- fazendo TV no Brasil*. Rio de Janeiro: Zahar.

Fiske, J. (1987) *Television culture*. London; New York: Methuen.

Harari, Y. (2015) *Sapiens, uma breve história da humanidade*. Porto Alegre: L&PM.

Johnson, C. (2019) *Online TV*. London: Routledge.

Kogut, P. (2020) O amor dos brasileiros pela telenovela só cresce. *O Globo*. [online] Available at: https://kogut.oglobo.globo.com/noticias-da-tv/critica/noticia/2020/07/o-amor-do-brasileiro-pelas-novelas-so-cresce.html. Accessed: 26 July 2020.

Miller, D. (2013) Future Identities: Changing Identities in the UK – The Next 10 Years. *Gov.UK*. [online] Available at: www.gov.uk/government/publications/future-identities-changing-identities-in-the-uk. Accessed: 25 July 2020.

Muanis, F. (2018) *A imagem televisiva- autorreferência, temporalidade, imersão.* Curitiba: Appris Editora.

Shirky, C. (2010) *Cognitive surplus: How technology makes consumers into collaborators.* [ebook] New York: Penguin. Accessed: 10 June 2010.

Svartman, R. and Nogueira, E. (2018) Interview with Eneida Nogueira.

Svartman, R. and Sánchez, M. (2018) Interview with Mirian de Icaza Sánchez.

Index

Note: Page numbers in *italics* indicate a figure on the corresponding page.